Race to the Finish

Series Editor
PAUL RABINOW

A list of titles in the series appears at the back of the book

Race to the Finish

Identity and Governance in an Age of Genomics

Jenny Reardon

PRINCETON UNIVERSITY PRESS
PRINCETON AND OXFORD

Published by Princeton University Press, 41 William Street,
Princeton, New Jersey 08540

In the United Kingdom: Princeton University Press,
3 Market Place, Woodstock, Oxfordshire OX20 1SY

Library of Congress Cataloging-in-Publication Data

Reardon, Jennifer, 1972–
Race to the finish : identity and governance in an age of genomics /
Jenny Reardon.
p. cm. — (In-formation series)
Includes bibliographical references and index.
ISBN 0-691-11856-6 (cl : alk. paper) —
ISBN 0-691-11857-4 (pbk. : alk. paper)
1. Human population genetics—Social aspects. 2. Human Genome
Project. I. Title. II. Series.
QH431.R248 2004
306.4′5—dc22 2004050569

British Library Cataloging-in-Publication Data is available

This book has been composed in Sabon with Futura display

Printed on acid-free paper.∞

pup.princeton.edu

Printed in the United States of America

10 9 8 7 6 5 4 3 2 1

For my brother, J. Derek Reardon (1971–2002)

Contents

Acknowledgments ix

Chapter 1: Introduction 1

Chapter 2: Post–World War II Expert Discourses on Race 17

Chapter 3: In the Legacy of Darwin 45

Chapter 4: Diversity Meets Anthropology 74

Chapter 5: Group Consent and the Informed, Volitional Subject 98

Chapter 6: Discourses of Participation 126

Chapter 7: Conclusion 157

Appendix A: Methodological Appendix 169

Appendix B: Code for Interviews 173

Appendix C: Human Genome Diversity Project Time Line 175

Notes 179

Bibliography 211

Index 229

Acknowledgments

I could not have imagined a finer intellectual and personal journey than the travels that gave rise to this book. I am indebted to the many people and institutions who made it possible.

My first words of thanks must go to the researchers, activists, community members, and government officials who over the years helped me to understand the many complex and vital issues at stake in studying human genetic diversity. Participants in the Diversity Project debates spent countless hours with me, often fitting me into very busy schedules. Many read portions of the manuscript. I am deeply indebted to all for sharing their recollections, ideas, and concerns, and for tolerating the formalities of the interviewing process. I hope this book will only be part of many continuing conversations.

As much as this project could not have been undertaken without the shared expertise and energies of those involved in the Diversity Project debates, it never could have been imagined without the support of my dissertation committee. I am grateful to Sheila Jasanoff for the careful attention she gave to this book at every stage in its development. Without her intellectual vision, rigor, and guidance it simply could not have been written. Stephen Hilgartner guided me through all aspects of researching and writing about contemporary science. His unflinching support made a very ambitious endeavor practical and possible. Evelynn Hammonds taught me to read for all the nuances that any critical study of race requires. Without her support, I would not have undertaken this project. Anna Marie Smith trained me to read difficult theoretical texts. Any rigor I bring to the reading of political and social theory I first learned from her.

Beyond my immediate committee members I am grateful to many other friends and colleagues who commented on drafts, discussed ideas, listened to talks, and in other ways improved the thought and writing of this book. Thank you to Jay Aronson, Roberta Bivins, Geoffrey Bowker, John Carson, Claudia Castaneda, Simon Cole, Arthur Daemmrich, Michael Dennis, Robert Doubleday, Jim Dratwa, Joe Dumit, Steve Epstein, Sarah Franklin, Sara Friedman, Alan Goodman, Herbert Gottweis, Donna Haraway, Cori Hayden, Stefan Helmreich, Evelyn Fox Keller, Ronald Kline, Andy Lakoff, Hannah Landecker, Marybeth Long-Martello, Michelle Murphy, Carl Pearson, Paul Rabinow, Sara Shostak, Stefan Sperling, Chris Sturr, Kaushik Sunderrajan, Mariachiara Tallacchini, Karen-Sue Taussig, Charis Thompson, Elly Truitt, Debbie Weinstein, and David Winickoff. Special thanks to Susan Conrad, Lisa Gannett, Rebecca Herzig, Kelly Joyce, and Jeanne Winner, who at key moments read and commented on chapters and provided advice on the revision process.

I also want to thank in particular my colleagues and students at Brown University and Duke University for buoying me these last two years as I revised the manuscript. I am particularly indebted to Priscilla Wald and Robert Cook-Deegan, who provided crucial moral and intellectual support in the final phases of the writing process. Anne Fausto-Sterling, Elizabeth Weed, and the extraordinary members of the 2002–2003 Seminar of the Pembroke Center for Teaching and Research on Women opened up new intellectual dimensions in my research and helped me to articulate my findings across disciplinary boundaries. Lundy Braun and Chris Amirault provided sound words of advice when they were most needed. My students, in particular Jordan Blumenthal, Brady Dunklee, Jackie Mahendra, Catherine Trimbur, and Sandy Wong, always brought the ideas alive in new ways and reminded me why, even on my longest and most exhausting days, I could not think of a better way to earn a living.

I presented early versions of this work in talks or seminars at Brown University; Cornell University; Harvard University; the International Society for History, Philosophy, and Social Study of Biology; the Max Planck Institute for the History of Science; the MIT/Harvard History of 'Race' in Science, Medicine, and Technology Workshop; the National Institute of Environmental Health Sciences; Rutgers University; the Science Studies Research Seminar at the Belfer Center for Science and International Affairs; the Society for Social Studies of Science; the University of Kansas; the University of Manchester; the University of Lancaster; and the Virginia Polytechnic Institute. The comments and insights of these audiences greatly improved this work.

This work would not have been possible without numerous sources of financial and institutional support. Resources that enabled the researching and writing of this book came from Cornell University's Department of

Science and Technology Studies; a National Science Foundation predoctoral fellowship and Dissertation Improvement Grant (SBR-9818409); a travel grant from the Cornell Ethical, Legal and Ethical Issues Program; an NSF training grant entitled "Reframing Rights: Constitutional Implications of Technological Change"; a predoctoral and postdoctoral fellowship in Science, Technology and Public Policy at Harvard University's Belfer Center for Science and International Affairs; and a postdoctoral fellowship in genome ethics, law, and policy at Duke University's Institute for Genome Sciences and Policy. Lillian Isaaks, Debbie VanGalder, Judy Yonkin, and Marta Weiner provided crucial institutional help. At Princeton University Press I thank the In-Formation series editor Paul Rabinow for his critical vision; senior editor Mary Murrell for making the writing of a first book a remarkably smooth process; and copy editor Marsha Kunin for her abundant patience and many insights.

The artwork for the cover was done by Niki Lee. I am grateful to her for allowing me to borrow Appaloosa, who, alert and aware, embodies the spirit with which I hope it might be possible to approach genomic decodings of human differences.

Finishing this book—from cover to appendices—has given me the chance to appreciate just how long and involved the journey has been. I could not have undertaken it without the unparalleled love and support of many friends and family members. I want to thank in particular Reid Goméz, for believing in me as a writer and always pushing me to say more; Gerri Jones, for her sage wisdom and sense of adventure that sustained me throughout; Angela Moore, for inspiring me over the decades, and for encouraging me to develop the artistic dimensions of my work; and Kristin Buchholz, Susan DeLay, Becca Gutman, Saul Halfon, and Mary Klayder, for being there at the most critical times with unbridled love and support. Finally, my love and gratitude goes to Susan Conrad, who improved this book immeasurably, partly because she never let me forget what really mattered.

My parents deserve special recognition. For decades they supported my many endeavors—whether it was setting up a laboratory in their basement, or traveling to horse farms. They always believed in me, and my ability to become anything—a jockey, a scientist, or a scholar in a field I never could quite describe, for a job I never could quite specify.

Special words of thanks also go to Rebecca Herzig. In life, if you are lucky, you meet someone who helps you to understand what you want to do, and then gives you the courage and support needed to do it. I have been so lucky. Rebecca, thank you for everything.

Finally, my heart extends to all my colleagues and friends who helped my family and me through my brother's illness and death. It is hard to even know how to begin to acknowledge the tragedy and trauma of such a loss, and the support that I received, and continue to receive, that helps me through.

I dedicate this work to you, Derek. Our time together on this earth was cut short. However, our collaboration continues. I will always share with you a commitment to living life to its fullest, and to throwing one fine party.

■ ■ ■

Sections of Chapter 5 represent revised parts of an essay that first appeared under the title "The Human Genome Diversity Project: A Case Study in Coproduction" in *Social Studies of Science* 31[3] (2001): 357–88.

Race to the Finish

Chapter 1
Introduction

By all accounts, no one expected it.

In the summer of 1991, leading population geneticists and evolutionary biologists from the United States proposed a project to sample and archive the world's human genetic diversity (Cavalli-Sforza et al., 1991).[1] The proposed survey, they argued, promised "enormous leaps" in our understanding of "who we are as a species and how we came to be" (ibid., 491; Human Genome Diversity Project 1992a,1). To realize these promised advances in knowledge, proponents urged the scientific community to act swiftly. Social changes that facilitated the mixing of populations, they warned, threatened the identity of groups of greatest importance for understanding human evolutionary history—"isolated indigenous populations" (Cavalli-Sforza et al., 1991). To unravel the mysteries of human origins and migrations, these valuable gene pools would need to be sampled before they "vanished" (ibid.). The resulting time pressure, and the tens of millions of dollars it would take to conduct the survey, posed substantial challenges. Proponents recognized these constraints. It crossed nobody's mind that the project might one day be accused of inventing a new form of colonialism.

Initially, the proposal captured the imaginations of leaders in the human genomics community worldwide. The Human Genome Organization (HUGO), an international body responsible for coordinating activities within the Human Genome Project, formed a committee to investigate how to carry the initiative forward. The National Science Foundation (NSF), the National Human Genome Research Center (NHGRC), the National Institute of General Medical Sciences (NIGMS) and the Department of Energy (DOE) provided funds for three planning workshops. With this support in place, by the end of 1992 organizers had every reason to believe that what had become known as the Human Genome Diversity Project would begin operation by 1994.[2]

Their expectations, however, were disappointed. Far from winning support, in a series of events that many organizers have found inexplicable and even bizarre, the Diversity Project became the target of vociferous outrage and opposition shortly after the initiative's second planning workshop in October 1992.[3] In May 1993 some physical anthropologists accused the initiative of using twenty-first-century technology to propagate the concepts of nineteenth-century racist biology (Lewin 1993). In June of that year, indigenous leaders from fourteen United Nations member states drafted a declaration calling for an immediate halt to the initiative. In July the Third World Network charged the Project with violating the human rights of indigenous peoples by turning them into objects of scientific research and "material for patenting" (Native-L 1993a). And in December the World Congress of Indigenous Peoples dubbed the initiative the "Vampire Project," a project more interested in collecting the blood of indigenous peoples than in their well-being (Indigenous Peoples Council on Biocolonialism 1998). By 1998 over a hundred groups advocating for the rights of tribes in the United States and indigenous groups worldwide had signed declarations condemning the Project (ibid.).

In the aftermath of these events, the puzzle for many scientists, ethicists, and government officials who seek to study human genetic differences is how this seemingly beneficent and well-intentioned initiative came to be so stigmatized.[4] The Project's leaders included some of biology's most respected, socially conscious scientists—scientists who had devoted significant energy over many decades to fighting racism and promoting human rights. Mary-Claire King, a medical and population geneticist, used genetic techniques to assist the Abuelas de Plaza de Mayo (Grandmothers of the May Plaza) in their effort to identify grandchildren kidnapped during Argentina's Dirty War.[5] Luca Cavalli-Sforza, a human population geneticist, debated William Shockley, a Stanford physicist who called for the sterilization of women from "inferior races," during the race and IQ debates of the 1970s. Robert Cook-Deegan, a physician and geneticist, worked for Physicians for Human Rights. These were not self-seeking researchers who sought to extract the blood of indigenous peoples for the sake of financial and political gain. They were scientists who sincerely hoped to create a project that would deepen the stores of human knowledge while fighting racism and countering Eurocentrism (Bowcock and Cavalli-Sforza 1991, Cavalli-Sforza 1994). It would be historically inaccurate, and morally insensitive, to understand the Diversity Project as an extension of older racist practices by labeling the initiative the product of white scientists wielding the power of science to objectify and exploit marginalized groups. The story of the Project is more complicated. It raises questions that cannot be resolved so easily.

I argue in this book that, far from being a straightforward story about the powerful exploiting the powerless, the Diversity Project debates raise fundamental questions about how to understand the very constitution of power and its relationship to science in an age when scientific claims about the human—in particular, its genomes—increasingly influence decisions about how humans should regulate and conduct their lives. Dominant analytic frameworks in the social sciences assume power distorts science and the work of scientists—leading them, for example, to produce racist ideologies. Science, in reverse, is the antidote to power; it produces truth that counters ideologies. In the case of the Diversity Project, however, this understanding of the oppositions between science and power, or truth and ideology, proved inadequate. Claims that the Project would lead to the end of racism by producing reliable scientific knowledge were just as unconvincing as some of the critics' claims that the Project would propagate racism and colonialism by exploiting the genes of indigenous peoples.

In order to understand the Diversity Project debates, a different understanding of science and its relation to power is needed. In place of a framework that casts science and power as already-formed entities that oppose each other, the simultaneous emergence of novel forms of knowing, and of governing the human, evident in this initiative, challenges us to find conceptual tools that will draw into view the ways in which knowledge and power form together. The Diversity Project raised fundamental questions about how to characterize human genetic diversity for the purpose of understanding human evolution and history. Yet, these questions about how to order and classify an aspect of nature to advance human understanding proved inseparable from an allied set of questions about how to organize human differences for the purposes of creating credible and legitimate systems of governance. The conceptions of science and power upon which many Project organizers relied did not bring these entanglements into sharp focus. Thus, organizers were continually caught off guard when questions about power—for example, questions about how to make authoritative claims about human diversity—turned out to be embedded in what they viewed as merely a scientific, humanistic, and anti-racist endeavor to understand the history and evolution of the human species.

Although the Diversity Project has ceased to move forward in its original form, the contentious questions it raised endure in their importance. Human genetic-variation research now tops the agendas of both private and public research institutions. In October 2002, the National Human

Genome Research Institute (NHGRI) announced the launching of a $100 million public-private effort to map human genetic variation, the International Haplotype Map (HapMap) Project (*http://genome.gov/10005336*). Countless other initiatives speckle the life sciences landscape as researchers interested not just in human evolution, but also in medicine and public health, seek to understand the human—its diseases, health, and potential—using the new powerful tools of the genomic revolution. Far from transcending the problems raised by the Diversity Project, these current efforts have only generated similar troubling questions (Couzin 2002).[6] By returning to the Diversity Project debates, this book seeks to bring into view and clarify the underlying contestations over the nature of knowledge, power, and expertise that were at stake in this effort to catalog human genetic diversity, and that continue to create discomfort today.

Race, Expertise, and Power

At the center of these contestations is a broader struggle over the meaning of race in science, medicine, and the modern state. As the last millennium ended, efforts to use racial categories in biomedical research and public health generated fundamental questions. Is race an obsolete concept that should be left behind by the operations of liberal democratic societies and scientific and medical research, insofar as this is possible, as some cultural and scientific critics have argued (Appiah 1990, Freeman 1998, Gilroy 2000, Wilson et al, 2001)? Or should race be understood as a positive category that designates both cultural and national belonging, and contributes to public health efforts to reduce the burden of disease (Du Bois 1961 [1903], Cruse 1968, NIH 1994, Risch 2002)? Can past misconduct and inequities in government, including the provision of health services, be overcome by transcending the concept of race, or, conversely, must this concept be actively employed to overcome those racial structures that continue to oppress?

Diversity Project organizers found themselves in a peculiarly ambiguous and paradoxical position with respect to these questions. On the one hand, they claimed that the Diversity Project would help "combat the scourge of racism" by demonstrating that "there is no absolute 'purity'" and no "documented biological superiority of any race, however defined" (Cavalli-Sforza 1994, 1, 10). Rather than promote racial division, Project organizers promised the initiative would demonstrate "humanity's diversity and its deep and underlying unity" (Cavalli-Sforza, 1994, 1). As one step toward these goals, one of the Project's main scientific leaders, the human population geneticist Luca Cavalli-Sforza, advocated abandoning the category of 'race' in favor of the categories 'group' and 'population' (Cavalli-Sforza 1994, 11).*

Yet, as we will see, at the same time that proposers of the Project disavowed the use of the category of race, they found themselves being accused by many of reinscribing old racial categories, and even of being racist. Some physical anthropologists argued that the Project employed categories that carried forward notions of racial purity (Lewin 1993, Marks 1995). Numerous indigenous rights groups charged Diversity Project organizers with continuing a long tradition of the West's use of racial science to justify its exploitation of the powerless (Mead 1996, Indigenous Peoples Council on Biocolonialism 1998). In an ironic turn, in the face of these critiques, some Project organizers began to explicitly employ racial categories. Representing what appeared to be a turnaround from the earlier disavowal of race, some leaders of the initiative now argued that the Project would include the genomes of African Americans and other "major ethnic groups," and in this way would serve as an "affirmative action" response to the Human Genome Project (Weiss 1993).

As the chapters that follow illustrate, underlying these debates about race were fundamental questions about how knowledge of the human should be produced in a genomic age, and who possesses the expertise needed to participate in this pursuit. How could human beings come to know their own species—its history and evolution—in an age when novel technologies enabled a purportedly molecular vision of human existence? What role, if any, could studies of human genetic differences play? How should such studies be designed? Which of the human sciences, if any, could provide the organizing concepts and methods? Human population genetics? Physical anthropology? Cultural anthropology? All of them?

These questions about the constitution of the right kind of knowledge were connected to questions about the nature of power. As Michel Foucault demonstrated through his studies of madness, the clinic, and the prison, the human sciences play central roles in constituting techniques and procedures for directing human behavior in the modern epoch (Foucault 1973, 1975, 1976). This modern age witnessed the entanglement of rules that govern what can count as knowledge with rules that determine which human lives can be lived. The result was the emergence of a new kind of power, what Foucault named *biopower* (ibid., 1976).

* In this book I use single quotation marks to indicate categories (like 'race,' 'population,' and 'group') and objects of study (like 'human genetic diversity'). To keep the use of quotation marks to a minimum, I often write out "the concept of" or "the category of" instead of using quotation marks. When introducing new concepts and categories I might use both methods (for example, in this case, "the categories of 'group' and 'population' "). Double quotation marks are reserved for direct quotes and in place of "so-called."

This close relationship between the rules that structure power and knowledge (a relationship Foucault highlighted through use of the contraction *power-knowledge*) was tellingly revealed in the attempts to plan the Diversity Project. Not surprisingly, a scientific endeavor that promised "enormous leaps in our grasps of human origins, evolution, prehistory, and potential" provoked questions about the role the initiative might play in producing human subjects who could be governed in novel, and potentially oppressive, ways (Cavalli-Sforza et al., 1991, 491). Would Diversity Project organizers' seemingly benevolent efforts to include groups in the design and the regulation of the sampling initiative facilitate their inclusion in an ethical manner, or would these efforts only threaten the sovereignty rights of tribes in the United States and marginalize other minority groups? Would the sampling and potential patenting of indigenous DNA bring benefits to indigenous groups through royalty sharing and medical advances, or would it merely enable their further objectification and exploitation? And who, if anyone, could speak for tribes and major ethnic groups in the United States, and indigenous groups across the globe, on these novel issues raised by a global survey of human genetic diversity? As these questions would make clear, at stake in the struggles over the Diversity Project were not only the validation of a research project, but also the resolution of some of the central social debates in an age defined by the emergence of genomics, globalization, deepening tensions between the (global) North and South, and a renewed struggle over the status and definition of race in the United States.

Co-Production: A Framework for Studying Emergence

To address these connections more directly, this book tells the story of the Project from within an analytic framework in science and technology studies that seeks to understand how scientific knowledge and social order are produced simultaneously—or, in a word, co-produced (Jasanoff 2004).[7] By linking critical studies of scientific knowledge and analyses of political institutions and social structures, I along with other scholars working within this analytic framework am attempting to clarify the ways in which bringing technoscientific phenomena or objects into being (objects, for instance, like 'human genetic diversity') require the simultaneous production of scientific ideas and practices *and* other social practices—such as norms of ethical research and credible systems of governance—that support them.

The resulting fine-grained portrayals of the mutual constitution of natural and political order have done much to undo grand narratives that have either celebrated science as an ideal polity (Merton 1973), or condemned it for reinscribing hegemonic and oppressive political orders (Habermas 1975). In the place of totalizing broad-stroked accounts, we have gained richer, more useful pictures of science and technology (S&T) that neither simply sanctify nor condemn, but rather bring to light the locally contingent and often ambivalent roles the creation of S&T plays in the ongoing human struggle to produce societal arrangements in which humane and meaningful lives can be lived.

On a terrain prone to polarization, the co-production idiom is a particularly valuable resource for countering analyses that too easily celebrate enlightened, objective studies of human genetic diversity, or too readily dismiss them as ideological and racist. It provides a framework that enables us to see that all efforts to organize such studies necessarily entail the production of both conceptual order and societal interests, and that the two domains are inextricably linked—indeed, inseparable. Such critical vision requires that the analyst give causal primacy to neither 'society' nor 'science,' but rather engage in a "symmetrical probing of the constitutive elements" of both (Jasanoff 2004). The result are accounts that resist both technological and social determinism, and easy pronouncements of right and wrong (Rabinow 1999).

The co-production framework is also appropriate for this study because of its utility for studying emergent phenomena. It is at the point of emergence, when actors are deciding how to recognize, name, investigate, and interpret new objects, that one can most easily view the ways in which scientific ideas and practices and societal arrangements come into being together (Jasanoff 2004, Daston 2000, Latour 1993). This is especially the case when the epistemological and normative implications of the emerging object are contested, and when the effort to establish its legitimacy and meaning spans multiple cultural contexts (Jasanoff 2004). All these conditions hold for 'human genetic diversity,' making it an especially rewarding site for analysis.

Scholarship guided by this analytic framework runs against dominant ideas about science in the academy and in society that conventionally have been linked to the Enlightenment. Enlightenment thinkers such as Voltaire, Rousseau, and Thomas Jefferson believed that through science and reason "men" could discover universal truths about Nature. It was on the basis of these truths, they argued, that man could recover from the blinding effects of dogmatic beliefs and uninterrogated traditions and achieve the enlightened stance required to build good and just governments. Later, Marxists would hail "scientific" thinking about political economy and the class struggle as a critical tool for piercing the veil of ideology.[8] In other words, science and reason, they believed, could be used to work against the corrupting influences of power divorced from truth.

Organizers of the Diversity Project shared this Enlightenment vision. They believed that their Project would generate scientific data that could oppose racist ideologies in society. Specifically, the first proposers of the Diversity Project joined critical race theorists, American historians, and cultural and biological anthropologists in their belief that population genetics demonstrated the biological meaninglessness of socially meaningful racial categories (Cavalli-Sforza 1989, 1994). Human population genetics, they accordingly argued, was a powerful antidote to the ideologies propagated by the use of these unscientific categories. Data generated by this discipline would lead to the final demise of the category of race in science and its replacement by the scientifically rigorous category population (Cavalli-Sforza 1994).

This analysis of race as an obsolete concept in the life sciences illustrates what might be called a debunking critique of ideology. This form of critique seeks to reveal hidden connections between discourse and social power.[9] The goal is to problematize the truth claims of a given discourse by demonstrating that these claims are the product of dominant, often reprehensible, social interests—and thus constitute ideology.[10] As noted above, this mode of critique—one that relies upon a clear distinction between power and knowledge—did not prove effective in the case of the Diversity Project. It provided explanations that no one found satisfactory.

Rather than dwelling on the opposition of truth on the one hand and power on the other, co-productionist work demonstrates that scientific knowledge and political order come into being together. Thus, government officials, policy makers, and academics cannot simply turn to scientific research (such as human genetic diversity research) for independent and objective answers to social problems, such as racism; nor can these social actors govern without the aid of science (or systematic knowledge-seeking), for scientific knowledge and ethical and political decisions about human diversity can only be made together. Human genetic diversity simply cannot become an object of study absent social and moral choices about what we want to know and who we want to become. Moral and social choices about the directions of human inquiry are not possible without cognitive frameworks that frame the human and its possible variations.

The story of the Diversity Project, I argue, can be more effectively told in this idiom of simultaneous emergence. This language allows one to step back from uncritical pronouncements about knowledge or assertions about power in order to ask a prior set of questions about how each is implicated in the formation of the other. To draw these processes of formation into focus, in this book I ask specifically about how categories used to classify human diversity in nature and those used to order relevant aspects of social practice shape, entail, and refer to each other: How do societal arrangements affect the kinds of categories that scientists can use to characterize human diversity? How do these categories in turn "loop back" to produce new societal arrangements (Hacking 1999)? What deliberative engagements between scientists, policy makers, research subjects, and citizens enable a socially meaningful biological category like population or race to work? As the chapters that follow demonstrate, formulating and answering these questions renders visible the micro-processes by which social life and cognitive understandings gain form and meaning together. The result is a rich, empirically grounded account of the entangled processes through which knowledge and social order sustain each other in contemporary societies.

Such an account promises to bring into view the tensions and paradoxes surrounding race raised by the Diversity Project. Just as co-productionist studies provide a way out of dualistic modes of understanding science (either it produces truth, or it is distorted to produce ideology), so too they provide a way out of the dichotomy that is prevalent in current analyses of race in the academy. Most conventional accounts portray race as either a valid category of research that can help produce legitimate knowledge (Dobzhanksy 1937 [1951], UNESCO 1952a, Risch 2002, Burchard et al., 2003), or as a social construct (i.e., ideology) propagated by the powerful (Gates 1986, Fields 1992, Cooper 2003). By adopting a co-productionist approach, this study demonstrates that race defies simple categorization as either the reflection of scientific truth or social ideology. As the chapters that follow demonstrate, a socially important category like race is likely to generate scientific attention. What makes this ordering tool of interest to scientists is precisely what makes it of interest to the law, criminal justice system, and institutions of higher learning—through centuries of use it has become a tool that draws into focus differences between humans deemed meaningful. At the same time, use of a socially meaningful category like race can never be refined so that it acts only to elucidate natural reality. As much as biologists have tried over the last several decades to constrict race to apolitical scientific purposes, the use of race is never neutral.[11] It is always tied to questions with political and social salience. Some of these questions—about, for example, resource allocation—are relatively benign. Others, such as questions about the creation of new forms of racism, are of great political significance.

Rather than a category that respects any demarcation between the scientific and social realm, race has traveled vigorously and often across the boundaries of science and society, reality and ideology, throughout the twentieth century. In the process, it has been stabilized and destabilized, made and remade. Analyzing the Diversity Project debates using the co-productionist framework allows us to draw into focus the simultaneously scientific and social dimensions of contemporary attempts to define and construct this category, and promises to create new insights into the dilemmas of categorizing human difference that confront us in the genomic age.

At the same time, analyzing race in this manner promises new understandings of the processes by which natural and social order come into being together. In many important senses, struggles over the meaning of race and racism are at once contestations over who has the expertise needed to represent the truth about human diversity in nature, and contestations over who has the authority to create just representations of human diversity in society. This is particularly true in the United States where debates over how to define race and racism are perennially at the center of battles over the nature of truth and justice. Especially in the post–World War II era, to be deemed a racist was to risk losing one's status as a speaker of truth *and* one's authority as a credible voice in society (Southern 1971, Baldwin 1998). As we will see in the chapters that follow, at stake in the contestations over what constituted racial categories, and the role—if any—these categories should play in ordering the Diversity Project were not just answers to fundamental questions about what will count as the truth about human nature and its diversity, but also answers to fundamental questions about who will count as legitimate members of particular societies. Who will be represented? Whose voices will be deemed authoritative or at least worth hearing? What structural effects will result? In short, focusing on the debates about race and racism sparked by the Diversity Project enables us to bring into view the processes by which knowledge and social order form together. This book argues that it is only by bringing these processes into view, and making them available for critical debate and understanding, that we will begin to make sense of and meaningfully address debates sparked by the Diversity Project.

Excavating the History of Race and Science

At the same time that this book is an interpretive project concerned with how objects (e.g., 'human genetic diversity') gain both scientific and broader social meanings, it is also necessarily a historical one. Objects do not come into being de novo. Rather, they are the products of long historical processes that embed past contestations and settlements (Daston 2000). Bringing these historical processes into view is vital to the task of understanding why some phenomena gain the material and intellectual support needed to persist, while others fade away.

This is especially true in the case of 'human genetic diversity.' As we will see, the efforts of Project organizers to render this phenomenon an object of sustained scientific inquiry raised historically entrenched questions that were in the deepest sense simultaneously scientific and social: What kinds of human diversities matter? Cultural or biological? The diversities of individuals or groups? If biological, how should the biological realm be defined? If groups, how should these groups be defined, by whom and for what purposes? In the interwar and post–World War II era, questions about the role that ideas and practices of race should play in resolving these questions have been at the center of debate. Is race, many have asked, one such grouping category that can be used to order human genetic diversity? Or is this term too confused and/or dangerous to be employed by scientists or any other actors in society?

Discussion and debate of these questions as they connected to the Diversity Project ostensibly began in the context of efforts to organize the Human Genome Project. In October 1990 the United States National Institutes of Health and Department of Energy authorized $3 billion to be spent over fifteen years to sequence a single human genome.[12] One of the main assumptions underlying this endeavor was that all human genomes were enough alike to create such a record. The initial proponents of the Human Genome Diversity Project challenged this assumption. As the human population geneticist and founding father of the Project, Luca Cavalli-Sforza, explained: "Each group has its differences and each person has differences. If we don't understand that diversity, we're missing a lot that's important" (Cavalli-Sforza quoted in Rensberger 1993). One consequence, he argued, would be the continuation of the "Eurocentric" bias of studies of human genetic diversity, studies that to date had "been made on Caucasoid samples for obvious reasons of expediency" (Cavalli-Sforza and Bowcock 1991, 491).

Yet the question about the value of studying diversity far predated the genomics era. During much of the first half of the twentieth century, it lay at the heart of a debate between experimental and population geneticists about how to study human evolution. Classical experimental geneticists assumed for the purposes of their research that all genomes were essentially the same. Based on this assumption of essential similarity, this group of scientists sought to discover the purportedly universal basic mechanisms that regulated and controlled all life. In contrast, prominent evolutionary biologists interested in genetics—most notably, the population geneticist Theodosius Dobzhansky—began their work and research from the assumption that all individuals are unique. For these scientists, diversity did not constitute deviation; rather, it was a normal and critical part of the natural world that provided the biological material upon which natural selection acted. Far from being a secondary concern, genetic diversity proved central to understanding evolution (Provine and Mayr 1982).

Within the group of scientists who agreed that genetic diversity would be a meaningful object of study, another debate emerged about how this diversity should be ordered. As we will see in the first part of this book, at the center of the debate was a struggle over whether and how to use race for the purposes of studying and interpreting human biological differences. This struggle first arose during the interwar period, sparked by the rise of eugenics and Nazi science. Following World War II, the use of the category of race in biology (and throughout the sciences) would be the subject of review by newly emerging international institutions. Most notably the United Nations Educational, Scientific and Cultural Organization (UNESCO) drafted a Statement on Race in 1950 that would be followed by a Second Statement published in 1952 (UNESCO 1952a).[13]

Most historians of biological beliefs about race hold that these UNESCO Statements on Race mark the beginning of the decline of race as a meaningful biological category, and its emergence as a sociological category. On their account, by mid-century scientists had lifted the veil of ideology that previously shrouded biological studies of human diversity. The legitimate science of population genetics eclipsed the ideological science of race biology; population replaced race as the category that biologists believed most usefully organized their analyses of human diversity (Stepan 1982, Barkan 1992).

Yet despite these histories, many, including the well-regarded founding father of population genetics, Dobzhansky, would continue to find race useful (Provine, et al. 1981). "Races," he argued in his classic *Genetics and the Origin of Species*, "may be defined as Mendelian populations of a species which differ in the frequencies of one or more genetic variants, gene alleles, or chromosomal structures." He in turn defined a "Mendelian population" as "a reproductive community of individuals which share in a common gene pool" (Dobzhansky, 1951[1937], 138, 15). Dobzhansky argued that anthropologists' failure to recognize that races were Mendelian populations led to their "endless disagreement" and confusion about the meaning of this term (ibid., 140). Debates on this score would continue between and among population geneticists and physical anthropologists in the coming decades.[14]

In many ways, these debates among and between geneticists and anthropologists can be seen as precursors to the Diversity Project debates. Although they were not as drawn out or as formal as the later debates, they did involve not only the carving out of a conceptual order, but also the creation of a social space that could support research on the formation of human races. This effort took place at a time when segregation and race-based lynchings in the American South, eugenics movements, and, above all, World War II had sparked fundamental critiques of purportedly scientific studies of race. In this new environment, it would no longer be permissible to conduct research that might be associated with Nazi race science. Even the term race had started to become taboo. However, far from leading to the end of the history of race in science, as many historians on anthropology and biology have claimed, this new environment fostered efforts by geneticists and physical anthropologists to carve out an expert space in which their studies of human diversity and race formation could continue.

To date, however, scholars have largely overlooked these continuing debates. By their account, as noted above, uses of race in science ended at mid-century once progressive scientists revealed the concept's ideological underbelly. This rendering of the history, however, is too simple. Uses of racial categories in science did not come to an end following World War II. To the contrary, scientific debates about race proved just as persistent and contentious as did the parallel social debates. To understand the controversies surrounding the Diversity Project, these debates must be unearthed from the archival records. I undertake this task in the first section of this book.

My goal in conducting this historical analysis, however, is not merely to excavate scientists' continuing debates about the proper definition and uses of race following World War II, debates that shaped the terrain upon which Diversity Project organizers would attempt to build their initiative. Rather, in keeping with the co-production idiom, my aim is to reveal the inextricable links between these debates about race, and broader social debates about the role this category should play in classifying human differences in society. As in the sciences, in post–World War II societies more broadly, debates about how to understand the human encountered tensions between discourses of sameness (most notably, universal humanism), and discourses of difference (race, ethnicity, and nationhood). Discourses of sameness and unity, such as universal humanism, gained strength. Many viewed these discourses as antidotes to the logics of differentiation that led to the slaughter of millions of innocent lives during World War II. Instead of a world divided into superior and inferior individuals and groups, universal humanism imagined the "united family of man" (Haraway 1989, 198). This doctrine became embodied in institutions such as the new international organization, the United Nations, and official documents, such as the Universal Declaration of Human Rights (United Nations 1949 [1948]).

Simultaneously, discourses of difference persisted. Claims about racial and ethnic difference, for example, continued to play primary roles in the reconstruction and maintenance of national and global political orders. Reminiscent of debates about diversity in the natural order, however, efforts to define race for the purposes of creating social order raised many questions. How should race be determined? For what purposes? How, if at all, does race differ from ethnicity? In the wake of the Civil Rights movement and the adoption of "affirmative action" policies in the United

States, these struggles over the determination of racial and ethnic differences proved to be as much about access to social goods as they were about the denial of these social goods to certain groups. Some discourses of race and ethnicity helped to articulate policies designed to open up education and job opportunities for minorities, while others continued to undergird discriminatory practices (Executive Order 10925: 1961).

These changes were connected to a broader set of political transformations that accompanied the emergence of identity politics in the 1960s. Social and political theorists have celebrated the rise of social movements constructed around claims about identity. This strategic use of identity, they have argued, represented a departure from old models of identity formation in which the state classified and identified people (Melucci 1989). Race in this new era emerged as a powerful resource used by citizens to build civic identities that could be mobilized to make demands against the state.

In order to make sense of the Diversity Project debates, the initiative must be understood in the context of these broader social struggles and transformations. Far from transcending the dilemmas and tensions generated by the liberatory and discriminatory effects of recognizing human differences, today studies of human genetic variation play central roles in negotiating them.[15] Some identify genomics with the liberatory trends of identity politics. In recent years, for example, scholars of biomedicine and genomics have observed that novel genomic and biomedical techniques and ideas enable new identity formations (Epstein 1996, Rabinow 1999, Kaufmann 2001).[16] Some argue this biologization of identity is fundamentally different from older reductionist and eugenic models (Rabinow 1999, 9).[17] For example, the anthropologist of science Paul Rabinow believes that the rigid, oppressive nineteenth-century categories (such as race) have been replaced by categories that are "inherently manipulable and reformable," and offer new possibilities for life (Rabinow 1999, 1996). Many have heralded the category of population brought into being by population genetics as one such category with liberatory potential (Haraway 1989, Stepan 1982).

Some also believe that defining racial and ethnic groups in genetic terms is a necessary component of progressive affirmative action policies. Specifically, in recent years scientists and public health officials have argued that genetic research on particular ethnic populations will help bring more minorities into science, as well as make it possible for the biomedical sciences to address the medical needs of minority communities (DNA Learning Center 1992, Nickens 1993, Pollack 2003). Some also hope genetics might offer a less contestable method for determining that an individual is "Native American," thus creating new possibilities for gaining access to the resources that come from tribal membership in the United States (Yona 2000).

But others view the rise of human genomics as less liberating. Since the early days of planning the Human Genome Project, many have expressed concern that this new burgeoning field of research could extend discriminatory practices into new realms—such as the provision of health and life insurance—and provide these practices with new, more powerful and impenetrable (i.e., scientific) sources of legitimacy (Billings 1992, Natowicz 1992, Juengst 1995). Others fear that genetic diversity research will lead to a new eugenics (Guerrero 1998). Still others fear that race-based discrimination based on the old biology might be replaced by discrimination based on the new group categories of population genetics—for example, characterizing 'demes' might lead to "demic discrimination" (Juengst 1995).[18]

Huge investments of venture capital into genomics have also spawned new fears about the commodification of life and the transformation of biology into a tool of global capitalism (Rural Advancement Foundation International 1993). Using the terms "biocolonialism" and "bioprospecting," scholars and activists alike have observed links between exploitative capitalist practices and the emergence of biological (in particular, genetic) diversity as a site of informational and commercial value (Mead 1996, Whitt 1998). For example, they point to the extraction of the non-Western world's plant resources as a critical element of colonial expansion since the sixteenth century (Kloppenburg 1988, Shiva 1997). In the contemporary era, many argue that pharmaceutical companies, in conjunction with government-sanctioned research, are extending these practices from the domain of plants to the domain of humans (Hayden 1998, Harry 1995, Rural Advancement Foundation International 1993).

Finally, claims about the distinctness of Native peoples have long been at the center of debates about the extent and legitimacy of their claims to sovereignty rights (Deloria 1985).[19] Some worry that genetic research will not serve to legitimate tribal membership claims, but rather will be used to undermine these claims by highlighting the similarities between Native peoples and other inhabitants of the Americas (Indigenous People Council on Biocolonialism 2000).

As Diversity Project organizers only too slowly began to recognize, the Project was entangled in these broader questions about the role genomics might play in constructing identity and novel forms of governance. To demonstrate this, I describe in the book's second half three moves that organizers made to respond to critics and stabilize the Project: (1) diversify the experts involved in the planning of the Project; (2) adapt existing tools in Western biomedical ethics to fit the "group" contexts in which the Project would be operating; (3) include "major ethnic groups" in the United States and indigenous groups worldwide in the design and conduct of research.

I demonstrate that each move proved unable to assuage critics as they respectively failed to engage with questions about how the Project's effort to order human diversity in nature for the purposes of scientific research would be inextricably tied to questions about how to classify human differences in society. Instead, these frameworks presumed the prior existence of groups in nature and society. Further, they presupposed the existence of the expertise needed to discern and represent these groups. Finally, they assumed that geneticists, anthropologists, and health workers working with populations possessed this expertise. Thus, they did not draw into sharp enough focus fundamental questions about the existence of groups in nature and society, the role of race in ordering and defining these groups, and the locus of expertise for answering these questions.

In short, these frameworks enabled too much to escape Project organizers' critical attention. Despite their best intentions, in responding to their critics, organizers often inadvertently acted to exclude too many people from the debates—both scientists who held different views, as well as people who had not yet been deemed experts in the official sense, but whose lives were being affected by the genomic revolution, and whose knowledge might have provided important insights into what it meant to interpret and define human diversity using the tools of scientific (genetic) experts. Consequently the debate remained too narrow. Focused primarily on the practical task of organizing a DNA sampling initiative, proponents rarely asked the more fundamental and, I would argue, more compelling and relevant questions generated by the genomic revolution, of which their Project would be a central part: What kind of human is brought into being via genomic analysis? What are this human's possible variations? Who can speak for humanness in this genomic age? Who will decide what kind of lives can be lived and not lived?

This book seeks to re-center scholarly attention upon these questions. In so doing, it provides an opportunity to reflect on what is new and what is old in the realm of the human (Foucault 1966). Rather than determining whether we have moved beyond the old, rigid, state-centered categories of race to new flexible categories for identifying and making sense of ourselves, or are poised, simply, to produce new tools for reinscribing old racial categories and systems of oppression, in the chapters that follow I seek to take a step back and clarify the fundamental issues that underlie contemporary debates about genomic studies of human diversity. My goal is to bring into critical view the heretofore unconscious processes through which human genomic research reconfigures both nature and society. My hope is to generate understandings that can lead to more reflective human futures.

Chapter 2

Post–World War II Expert Discourses on Race

According to prominent historical accounts, something called "the idea of race in science" emerged at the beginning of the nineteenth century, reached its pinnacle at the end of the nineteenth century, began to decline at the beginning of the twentieth, and then met its final demise after World War II (Stocking 1968, Stepan 1982, Barkan 1992). As do the theories of race whose decline they document, these accounts assume the existence of a stable and static entity whose rise and fall can be charted. Consequently, historians of race and science, and the critical race theorists whose work they inform, have generally overlooked transformations in the meaning and uses of race as a scientific category.

This chapter begins to fill this surprising gap in critical analyses of race, one that must be addressed if we are to understand the Diversity Project debates, and the contentious questions they raised about the definition and utility of race for contemporary genomic research. Specifically, the chapter demonstrates that the history of race and science did *not* include a period of enlightenment in the middle of the twentieth century in which scientists pierced the ideological veil of race to find the category wanting for material reality. Rather, the interwar period witnessed scientists rejecting only some uses of race. Concepts of race order and stabilize particular regimes of science and power, and as these rise and fall in importance, so do the concepts of race they embody. Accordingly, theories of race that underwrote Nazi racial hygiene declined in science once the legitimacy of the Nazi regime had been undermined. However, other theories of race continued to order biological studies of human diversity. Rather than dismissing race as mere ideology, following World War II students of human diversity sought to refine its definition and use so that it could advance knowledge without legitimating discrimination, which they regarded as a social act separate from scientific activity. For most, this meant reforming studies of race so that they reflected the most recent scientific advances—particularly advances in genetic analysis and statistical tools for studying variation in natural populations.

To see this more complex set of negotiations, and to bring into view those ideas and techniques that geneticists and physical anthropologists used to reconstruct racial categories following World War II, in this chapter, and throughout the book, I adopt a constructivist approach to science, knowledge, and expertise. Although adept at bringing to light the constructed character of claims about race when they perceive them to have social origins, critical theorists of race to date have not called into question the constructed nature of claims about race when they deem them the product of legitimate science. Perhaps the most striking and important case of this oversight is their embrace of the claim that gained media prominence in the mid-1990s: "scientists say race has no biological basis" (Hotz 1995, Flint 1995, Alvarado 1995). Rather than interrogating how such claims about the biological meaninglessness of race are made, and how they shape and are shaped by broader social, political, and technical contexts, many critical race theorists draw upon these claims to bolster their argument that race is mere ideology (Gates 1986, Appiah 1992, Higgimbotham 1992, Fields 1990, Gilroy 2000). Consequently, the constellation of different ideas and material practices that come to count as scientific statements about race, and the social, cultural, and technical contexts that enable or prevent their propagation, have largely escaped critical analysis.

To address this blind spot, race itself, I argue, must be treated as a historical object, an object constructed differently in different contexts (Scott 1992). In this chapter, and in subsequent ones, I seek to never use the formulation "the concept of race;" rather I use language that refers to the particular concept of race I wish to discuss. I never treat claims about race made by scientists as if they were true, but always interrogate the broader technical, social, and historical milieus in which they become viable and credible.

I adopt the same approach to claims about the distinction between truth and ideology. As we will see, most critical analyses of race in the academy draw upon an understanding of ideology that opposes it to science and knowledge, positioning the use of race in science on the side of ideology, and claims that race is biologically meaningless on the side of science and right knowledge. A critical look at the historical record, however, reveals that the line between ideology and science proved much less clear to scientists working during the years that preceeded and followed World War II, years associated with the decline and final demise of the idea of race in science. Indeed, as we will see, these scientists' struggles over how to use and define race precisely centered on the debate over what could distinguish a proper scientific approach to the study of race from misguided ideological approaches associated with the "Negro problem" in the United States and the "race problem" in Europe (Myrdal 1944, UNESCO 1952a). Building a defensible distinction proved central to their efforts to maintain credibility at a time when World War II combined with the mounting pressures of the Cold War had radically altered the meaning of race (Southern 1971, Baldwin 1988). By assuming a clear distinction between science and ideology, previous analysts have missed these crucial episodes of boundary work (Gieryn 1983).

To bring these episodes into view, in this chapter I describe the structure of the historical narratives that to date have obscured struggles over the nature of science and ideology, and debates about the proper definition and uses of race in science that followed the purported demise of the category following World War II. I then open up key aspects of these narratives, demonstrating that far from settled historical facts, lying barely beneath their surface were fundamental and contentious questions: What constitutes meaningful human variation? What role should human-constructed categories play in ordering this variation? What role should race, in particular, play? The chapter documents struggles to answer these more fundamental questions, struggles that to date have eluded critical historical analysis. It focuses in particular on contestations over the meaning and use of race sparked by the publication of the First UNESCO Statement on Race in 1950, and the introduction of the so-called cline concept for the study of human variation in the early 1960s. Debates sparked by both events would be rekindled forty years later with the proposal for the Diversity Project in 1991. To begin to understand the struggles surrounding the Diversity Project, these historically entrenched debates must be brought into view.

Histories of Race and Science

George Stocking's 1968 classic, *Race, Culture and Evolution*, offers a natural starting point for our analysis. Stocking, a historian of anthropology, began his inquiry into the history of anthropology by asking the question that the historian Oscar Handlin had asked over a decade earlier in *Race and Nationality in American Life*: "What Happened to Race?" He answered as follows:

> Although the twentieth century has suffered some of the most inhuman manifestations of racism, and the problems of race relations can hardly be said to have been solved, I suspect that the intellectual framework in which these issues are perceived has been permanently altered—at least in this sense, I am inclined to view "race" as a characteristically nineteenth-century phenomenon. (Stocking 1982 [1968], xxii, 270)

In *Race, Culture and Evolution*, Stocking set out to document the rise and fall of race from the "intellectual framework" of anthropology. To do this, he charted changes in the forces that anthropologists appealed to in their explanations of human differences: in the eighteenth century, environmental explanations appealed to forces of "civilization" (government, religion, language, customs, material culture); in the nineteenth century, physical differences moved to the forefront of anthropological analysis as race became the concept used to explain human differences; in the twentieth century, environmental explanations once again became dominant, and culture emerged as the preferred concept for making sense of human diversity (ibid., 17, 38).

Throughout, Stocking defined race as referring to "permanent inherited physical differences which distinguish human groups." Race, he argued, presumed the existence of unchanging types. Thus, as a scientific concept, it originated and throve in "the essentially static, nonevolutionary tradition of comparative anatomy" (ibid., 29–30). The concept of culture opposed race: culture referred to the environment whereas race referred to nature.

The construction of race as a sign of nature and culture as a sign of environment structures all of Stocking's historiography. For example, to explain the shift in the nineteenth century from theories of diversity to theories of race, Stocking cites Cuvier's changing explanation of "stupidity." In 1790 Cuvier argued that "stupidity" resulted from "lack of civilization," but by 1817 he had changed his explanation; instead of environmental forces, such as the lack of civilization, Cuvier now argued that "certain intrinsic causes" led to the slowing of human development. Stocking read this transformation in Cuvier's views as a sign of a broader shift rejecting environmental explanations in favor of biologically deterministic ones linked to race.[20] Europe's need to defend slavery when it came under attack at the end of the eighteenth century explained this shift: "The idea of race arose as a defensive ideology," he wrote, supported by "anti-egalitarians" (ibid., 36–38).

Many of Stocking's claims and concerns permeated subsequent historical accounts. Historians who followed him also wrote about "the idea of race" and agreed that this idea packaged humans into static types, that it arose in the nineteenth century as an ideology to defend slavery, and that it declined in the twentieth when it was replaced by more egalitarian and scientific concepts (in anthropology, culture; in biology, population). Nancy Stepan's *The Idea of Race in Science* provides a key example.

Like Stocking, Stepan argues that something called "the idea of race in science" arose in the nineteenth century.[21] Human races, she explains, became objects of systematic investigation by scientists at the end of the eighteenth century. In the first half of the nineteenth century, developments in geology, paleontology, and comparative anatomy led to the introduction of the idea that races formed a graded series, "with the Europeans at the top and the Negro invariably at the bottom." This gradation formed the cornerstone of the new "racial science" that would govern the study of human differences throughout the nineteenth century (Stepan 1982, ix, 6).

Stepan, like Stocking, argued that scientists used race to posit the existence of an "underlying essence or type" (Stepan 1982, xviii). Scientists discovered these types through measurements and quantification. The result "was to give a 'mental abstraction an independent reality,' to make real or 'reify' the idea of racial type when in fact the type was a *social construct* which scientists then treated as though it were in fact 'in nature' " (Stepan 1982, xviii; emphasis added). In Stepan's view, this reification lay at the heart of the errors of racial science.

This science, Stepan argued, did not exist for very long before ideas arose that eventually led to its decline. The second half of the nineteenth century witnessed the rise of Darwinian theories of evolution and natural variation (Stepan 1982, 47). These theories gave rise to what Stepan heralded as a "new science of human diversity" that opposed the study of static races with a study of natural variation in populations (Stepan 1982, ix). However, it would take nearly one hundred years before the new ideas of human diversity would be fully accepted. Typological notions of race proved to be deeply rooted "ideological" constructs (Stepan 1982, xx). Like Stocking, Stepan linked these ideas to the rise of the European slave trade, arguing that they acted to protect slavery when it came under attack at the end of the eighteenth century. Given the durability of this institution, she concluded, the the idea of race would not easily be displaced, not even by new science (Stepan 1982, xii).

To overcome these ideological forces, Stepan held that more than the "new science of human diversity" would be required.[22] Also needed would be the courageous efforts of "scientists uncommitted to the old racial thinking, often working on the outside of the discipline of anthropology" (Stepan 1982, 171). This group of reformers included Theodosius Dobzhansky, Ernst Mayr, J.B.S. Haldane, Sewall Wright, and others responsible for the "revolutionary works" of the "new, non-racial, populational, genetic science of human diversity" (Stepan 1982, 173–74).

The final prominent contribution to the history of "the idea of race and science" I consider is the cultural historian Elazar Barkan's book, *The Retreat of Scientific Racism*. Published a decade after Stepan's work, in this book Barkan argued that the "perception" that race was a scientific concept was a "legacy of the nineteenth century." During the second half of the nineteenth century and the first part of the twentieth century "typologies and hierarchies of race were presented as self-evidently appropriate . . . the inferiority of certain races was no more to be contested than the law of gravity to be regarded as immoral," Barkan wrote. This would remain the case, he observed, until after World War I, when scientists began to undermine the biological validity of race. It would take little more than a decade of critique before scientists would change their views dramatically. Most notably, on Barkan's account, scientists called into question typologies of race: "Among leading scientific circles in the United States and Britain, race typology as an element of causal cultural explanation became largely discredited, racial differentiation began to be limited to physical characteristics, and prejudicial action based on racial discrimination came to be viewed as racism" (Barkan 1992, 2–3).

The main reason for "the decline" of something called "the scientific idea of race," Barkan posited, was "a lack of epistemological foundation for racial classification, a lack which led to endless irresolvable inconsistencies and contradictions." Matching measurements of skulls and other bodily features to racial categories proved inconsistent, leading to a "classification quandary" that "made formal taxonomy impossible." Like Stepan, Barkan found that the category of race retained its "popular appeal," but that it fell out of favor in science. Drawing upon a "distinction between race as a scientific idea and as a social category," he too concluded that, as a scientific idea, race fell out of favor; as a social category, though, it remained in use (ibid., 3–4).

Throughout his account of "the retreat of scientific racism," Barkan assumed that a clear boundary could be drawn between the "factual" and the "ideological," "scientific data" and "cultural" and "social" bias."[23] The former represented a necessary but not sufficient condition for anti-racism. The latter led to "racial science."[24] Stepan would have concurred.

From the accounts of Stepan, Barkan, and Stocking we begin to see the outlines of the canonical narrative of the history of race and science: race emerged as a concept used by scientists in the nineteenth century, fell out of favor during the second two decades of the twentieth century, and then was replaced by studies of population and culture following World War II. Race was a typological concept that categorized humans into discrete units. Rooted in deeply entrenched racist ideologies, it could not be defeated by new scientific theories alone—the commitment of anti-racist, socially responsible scientists was also required: in biology, population genetics replaced the old typological racial science; in anthropology, a study of culture replaced the study of race (Barkan 1992, 210).

By the mid-1980s, this narrative had spread beyond observers of science to critical race theorists such as Henry Louis Gates, Kwame Anthony Appiah, Evelyn B. Higginbotham, and Barbara Fields (Gates 1986, Appiah 1992, Higgimbotham 1992, Fields 1990). Henry Louis Gates, in an introduction to a founding text of critical race theory, wrote in 1986: "Race, as a meaningful criterion within the biological sciences, has long been recognized to be a fiction." Citing Stepan, Gates argues that rather than proving "reports of reality," assertions of racial difference are metaphors that seek official sanction in science (Gates 1986, 4, 6). Gates and other critical race theorists turned to scientists and historians of science to legitimate their claims that race had no naturalistic meaning, and that biological constructions of race represented "ideological resistance to the truth" (Appiah 1992, 7).

This dominant narrative truncates history and removes from critical view a major site for the construction and reconstruction of race following World War II: the life sciences. As I will show in the remainder of this chapter, following the Second World War, scientists did not all agree that a concept of race no longer had any meaning in science. Instead, intense debates remained unresolved. To bring these debates into view, I will examine two key episodes: the publication of the first two UNESCO Statements on Race in 1950 and 1952, and the debate about race sparked by the physical anthropologist Frank Livingstone's introduction of the cline concept in the early 1960s. By examining these episodes through the lens of an analytic framework that does not simply treat the distinction between science and ideology as a conceptual tool for historical analysis, but rather draws into focus the work needed to maintain this distinction, a rich terrain of continuing struggles over the meaning and proper use of race comes into view.

The UNESCO Statements on Race

I begin my analysis with the publication of the 1950 UNESCO Statement on Race as many observers of the history of race and science cite the Statement as marking the end of "racial" (Stepan's language) and "racist" (Barkan's language) science. As Barkan argues, "The 1950 declaration was the first in an ongoing series of UNESCO's statements on the concept of race, and it displayed the environmental determinism at its peak. The reversal in the scientific credo on race since the 1920s had been completed" (Barkan 1992, 341). Similarly, in a recent volume on race, ethnicity, and nationalism, the editor Winston A. Van Horne writes: "[A]s the 'UNESCO Statement of 1950' calls out, there is indeed 'no biological reality to the concept of race' " (VanHorne 1997, 7).[25] As we will see below, however, these claims that the UNESCO Statements marked the end of the history of "the concept of race" in science do not match the historical record. Far from operating as a moment of closure, the publication of these documents ushered in an era of old and new debates about the use of race as an analytic category in science.

To understand these debates requires placing them in their broader political and historical contexts. In particular, they must be understood in the light of simultaneous efforts by scientists and policy bodies to address anxieties about the use of racial demarcations by national governments. These efforts began as early as the 1920s. During this decade, the U.S. Congress sought to restrict immigration to the United States—in particular, the immigration of Eastern and Southern Europeans and Asians. To aid this cause, the chairman of the House Committee on Immigration and Naturalization appointed Harry H. Laughlin, the administrative officer for Charles Davenport's Cold Spring Harbor eugenics laboratory, as "Expert Eugenical Agent" (Barkan 1992, 195). According to some historians of science, this appointment raised concerns among biologists that their science was being distorted for political ends (Ludmerer quoted by Barkan 1992, 197). However, others historians, such as Barkan, have contested this view, pointing out that the only American biologist to oppose immigration and Laughlin's pro-eugenic views in the 1920s was Herbert Spencer Jennings (Barkan 1997, Paul 1995).

By the 1930s, however, some American biologists and anthropologists had organized themselves to oppose the use of their research for what they perceived to be political ends. For example, seven leading Columbia University professors—including the noted geneticist Leslie Clarence Dunn and his colleague, one of the founders of population genetics, Theodosius Dobzhansky—founded the University Federation for Democracy and Intellectual Freedom (UFDIF). These scientists sought to combat Nazi theories of race through promoting intellectual freedom. According to the UFDIF, totalitarian political regimes like the Nazi regime propagated "misuse[s] of the term *race*" (Kuznik 1987, 201–207). Science, they argued, should be freed from such political distortions. As we will see, these calls to separate science from politics would continue through to the 1990s when controversies broke out about the potential entanglement of the Human Genome Diversity Project with racism.

The 1930s also proved a time of renewed attention to what became known as the "Negro problem." Although President Abraham Lincoln issued the Emancipation Proclamation in 1863, and Congress passed the first Civil Rights Act in 1866, the principle that liberty was inseparable from equality took much longer to root itself in U.S. society. By the end of the nineteenth century, the number of lynchings in the American South had exceeded three thousand, the voting rights of the Negro had been all but repealed, and segregation laws (later known as Jim Crow laws) had been passed by most Southern states.[26] As the legal historian Richard Kluger has argued, by the turn of the century nothing short of the "legalized degradation of the Negro in the form of a state-mandated caste system had begun" (Kluger 1975, 44–47, 68).

Segregation, disenfranchisement, and lynchings however, would not be considered significant social problems until many decades later. Indeed, it was only in 1937 that the Carnegie Corporation commissioned a study of "the Negro problem in America" (Myrdal 1962 [1944], xlvii).[27] This massive $300,000 undertaking, which employed seventy-five scholars, was portrayed at the time by both the director of the study, the Swedish sociologist Gunnar Myrdal, and by one of the nation's Negro leaders, W.E.B. DuBois, as a "bold" work that "proved America's basic soundness and strength." Some historians, however, have suggested that if one places the study in its broader political contexts, a different interpretation emerges. Instead of reflecting the strong moral fabric of the nation, the study reflected a need to constitute the *image* of a strong moral fabric for the purposes of establishing and maintaining the United States's position as a world power with imperial interests in Africa, Asia, and the Middle East. For example, David Southern, the best-known historian of the Myrdal study, quotes one "distinguished American" who explained: "Every time some race-baiter ill-treats some man in America he lessens the ability of America to lead the world to freedom." Indeed, Southern notes that Myrdal himself recognized America's "world prestige and future security would be related to how the Negro problem was handled" (Southern 1971, 3–4, 6).[28]

More recently, the critical race theorist Kate Baldwin and the sociologist John Skrentny have argued that efforts to create a colorblind society in the United States during the 1950s must be understood in the context of the Cold War (Baldwin 1998, Skrentny 2002). Soviet accusations that the United States's economic success depended upon the exploitation of colonized peoples created, in Baldwin's view, intense political pressure on the United States that led to a new way of talking about race. In this new language "blacks and whites were in solidarity against the real different folks, the Soviets" (Baldwin 1998, 107).[29]

It is in this broader historical and political context in which racism had emerged as a social problem (Gusfield 1981) that natural and social scientists convened to write the UNESCO Statements on Race. As we will see, this Statement attempted to position scientific research as an instrument of "rationality" that could fight the "ideology" of race prejudice. One significant problem remained. No consensus emerged on the fundamental issue of how rational scientific knowledge could be distinguished from the irrational forces of ideology.

THE FIRST UNESCO STATEMENT ON RACE

Drafters of the First UNESCO Statement left no doubt that the problem they sought to address was nothing less than the problem of ideology and the threat it posed to rational, enlightened democracies. Indeed, the UNESCO document in which the Statement first appeared began:

> Since the beginning of the nineteenth century, the racial problem has been growing in importance. A bare 30 years ago, Europeans could still regard race prejudice as a phenomenon that only affected areas on the margin of civilization, or continents other than their own. They suffered a sudden and rude awakening. . . . The virulence with which this *ideology* has made its appearance during the present century is one of the strangest and most disturbing phenomena of the great revolution of our time. Racial doctrine is the outcome of a fundamentally anti-rational system of thought and is in glaring conflict with the whole humanist tradition of our civilization [italics added]. (UNESCO 1952a, 5)

To oppose this ideological doctrine, UNESCO convened a group of "experts on race problems" whose charge was to disseminate "scientific facts designed to remove what is generally known as race prejudice" (Montagu 1972, x).[30] As the physical anthropologist and rapporteur for the meeting, Ashley Montagu described their charge:

> Race hatred and conflict thrive on scientifically false ideas and are nourished by ignorance. In order to show up these errors of facts and reasoning, to make widely known the conclusions reached in various branches of science, to combat racial propaganda, we must turn to the means and methods of education, science and culture, which are precisely the three domains in which Unesco's activities are exerted; it is on this threefold front that the battle against all forms of racism must be engaged. (UNESCO 1952a, 5–6)[31]

Racial hatred that led to the "great and terrible war," Montagu here argued, resulted from irrational ideas that opposed democratic principles of equality and mutual respect. Science, the institutionalized form of rationality, could best oppose this irrationality. UNESCO, the international institution most concerned with supporting science, had the duty to support this oppositional effort.[32]

Charged by the United Nations Economic and Social Council to fulfill this duty, the first UNESCO "panel of experts" convened at the UNESCO House in Paris in December 1949. This panel included members from Brazil, France, India, Mexico, New Zealand, the United Kingdom, and the United States. Additionally, it had been hoped that members would include both natural and social scientists, as the "study of man" was held to be the undertaking of both the natural and social sciences (UNESCO 1952a, 6). However, deaths and last-minute withdrawals meant that the final committee consisted mainly of sociologists. This would prove consequential.

Despite the "scanty representation of the biological sciences on the committee," the social scientists who remained agreed that "race had to be defined biologically" (UNESCO 1952a, 6–7). Race, they were firmly convinced, reflected evolutionary processes. These processes included "drift and random fixation of the material particles which control heredity (the genes), changes in the structure of these particles, hybridization and natural selection" (UNESCO 1952b, 76). Biologists' racial categorizations had been created to make sense of these *natural* processes.

Drafters of the first UNESCO Statement on Race went to great pains to distinguish this legitimate scientific use of race from ideological uses in society. At the time of the drafting of the First UNESCO Statement on Race, Nazi ideologies of racial inferiority and superiority were of concern to scientists and nonscientists alike. Although historians of the era differ over how these ideas should be studied and assessed, they do agree that the Nazis held the following views about race: races are pure; races are fixed and static; races are defined by innate biological differences reflected in psychological and cultural traits; race-crossing destroys racial integrity and leads to disharmonious physical results;[33] and there is a hierarchy of races.[34] Drafters of the First UNESCO Statement of Race set out to distinguish these theories of race, which they deemed the product of virulent political ideology, from those biological concepts of race employed by legitimate scientists. Their position can be summarized in the following five points.

First, geographic isolation (e.g., separation of human groups by oceans, deserts, and mountains) created races.[35] This explanation differed from the position ascribed to the Nazis in a critical way: instead of a directed force of nature (such as natural selection), it located the cause of racial differences in what were considered random environmental phenomena, such as separation of land masses and formation of deserts.

Second, scientists should use physical traits and not mental ones to interpret the formation of races. A document accompanying the Statement entitled *What is Race?* explained: "Races share a general tendency to produce certain physical traits." These traits included "hair, eyes, head shape, physique, etc." A table extending this list of physical traits included skin color, stature, face, nose, and body build (UNESCO 1952b, 6, 45). "Mental," "character," and "personality" traits, on the other hand, were found to have no place in racial categorizations: "Whatever classification the anthropologist makes of man, he never includes mental characteristics as part of those classifications" (Montagu 1972, 9). Two separate logics justified this claim. First, mental traits were seen as essential traits common to all humans; thus they were useless for purposes of discriminating among races. Second, the Statement argued that mental traits resulted from *environmental* forces and did not reflect the *natural* processes of evolution that give rise to racial differences (UNESCO 1952b, 29, 64).

Third, scientists define races as populations. Instead of pure types, races proved to be more like a statistician's population.[36] Physical traits might be expressed at a "higher frequency among members of the major groups to which they belong," but they always overlapped with other groups. Such traits could not be assessed through mere visual inspection; they required statistical analyses.[37] These analyses, the First UNESCO Statement argued, revealed that too many "border-line races and border-line individuals" existed to establish any single, agreed-upon, objective list of pure and distinct races (UNESCO 1952b, 64). As we will see later in the chapter, historians of race and science interpreted this move from typological to statistical analyses as indicating the replacement of race by population. Here, though, it is clear that population does not replace race; rather, races become populations.[38]

Fourth, races constitute dynamic, not static groups. They reflect the dynamic evolutionary processes that lead to changes over time (Montagu 1972, 8).

Fifth, there are no inferior or superior races. Authors of the UNESCO report explained that claims of inferiority and superiority depend on strictly segregating "meaningful" human traits (like intelligence and personality) into different groups. They challenged these claims on the grounds explicated above: meaningful human traits, they argued, have no connection to the evolutionary process and to race formation; racial traits do not segregate discretely into different groups. Thus there could be no hierarchy among races: the traits that one used to rank groups did not fall out along racial lines; the lines themselves were statistical, and not rigidly demarcated.[39]

The drafters of the First Statement hoped that these five points would serve to distinguish Nazi ideologies of race from what they viewed as scientific facts about race. Instead of the Nazi vision of inferior and superior races, these experts argued that the biological sciences supported a humanist vision of the unity of races. As the First UNESCO Statement proclaimed: "The unity of mankind from both the biological and social viewpoints is the main thing. To recognize this and to act accordingly is the first requirement of modern man" (UNESCO 1952b, 78).[40] The theme of human unity would again be emphasized forty years later with the proposal of the Diversity Project.

As these five points illustrate, drafters of the First UNESCO Statement on Race did not argue that scientists thought race was meaningless, but rather that scientists thought race was meaningless for particular purposes—namely, for the purposes of evaluating meaningful human traits and creating hierarchical distinctions among human beings. Their message was not that scientists did not use race, but rather that society should use scientists' understanding of race to guide their moral and social choices. Far from marking the exit of scientists from the social debate about race, this UNESCO Statement argued that society would be better off if scientists, with their expert knowledge, played a greater role.

One might imagine that geneticists and anthropologists (the scientists deemed experts on race) would have welcomed this 1950 UNESCO Statement. This was not the case. Indeed, far from endorsing the Statement, physical anthropologists and geneticists actively opposed it (UNESCO 1952a, 7; Haraway 1989, 201). The ensuing controversy revealed that far from contributing to the resolution of social anxieties by defining race scientifically, the 1950 UNESCO Statement on Race only served to stir up a more fundamental debate about what constituted a proper scientific definition, and what made it distinct from ideology.

THE SECOND UNESCO STATEMENT ON RACE

Following its publication on 18 July 1950, the First UNESCO Statement on Race received heavy criticism, especially from physical anthropologists and geneticists (UNESCO 1952a, 7; Haraway 1989, 201). For many of these scientists, the critical problem with the First UNESCO Statement was that while it advocated giving scientists a greater role in defining race in society, ironically it limited their ability to define race in their own research. Recall that the Statement barred mental and emotional traits from racial analysis. Physical anthropologists and geneticists argued that this could not be justified on "scientific" grounds and acted to irrationally restrict the scope of their research. Although they shared the 1950 Statement's drafters' goal of opposing Nazi race theory, they believed these drafters engaged in the same kind of distortion of science for ideological purposes practiced by the Nazis. The goal this time might have been to promote democracy, not totalitarianism, but it was no less misguided.

To address these criticisms, UNESCO convened a second committee of "experts on race," this time composed entirely of physical anthropologists and geneticists. The resulting Second UNESCO Statement on Race entitled, "Statement on the Nature of Race and Race Differences," written by physical anthropologists and geneticists, agreed with almost all the claims put forward by sociologists in the First Statement: races are derived from a common stock; there is no evidence for the existence of pure races; there are no inferior or superior races; races are deciphered inside scientific laboratories, not in society; the concept of race is merely a classification device used by scientists to study the evolutionary process; races are determined by several characteristics; races are not discrete groups defined by qualitative differences, but rather overlapping groups defined by quantitative differences.

There were, however, two critical changes. First, the physical anthropologists and geneticists of the 1951 Statement revised the sociologists' assertion that all meaningful human traits are "raceless," reasserting the possibility that traits pertaining to "intellectual and emotional response" could vary according to genetic differences between races. Specifically, it stated:

> It is possible, though not proved, that some types of innate capacity for intellectual and emotional responses are commoner in one human group than another, but it is certain that, within a single group, innate capacities vary as much as, if not more than, they do between different groups. (UNESCO 1952a, 13)

This claim accommodated both those who wanted to de-emphasize the importance of group-level differences in intellectual and emotional traits (by pointing to the importance of differences among individuals *within* a group) and those who wanted to hold onto a belief not uncommon among geneticists and physical anthropologists that group differences in mental and psychological traits did exist. As the geneticist Herman J. Muller explained:

> Since now there are these very abundant individual differences affecting psychological traits, it would be extremely strange if there were not also differences, in the frequencies of such genes, between one major race and another, in view of the fact that there are such major pronounced differences in the frequencies of genes affecting physically and chemically expressed traits. That would surely be the attitude of the great majority of geneticists. (UNESCO 1952a, 101)[41]

Other geneticists echoed this view. Kenneth Mather, a geneticist at the University of Birmingham, warned that the case against Nazi race theory is not strengthened by "playing down the possibility of statistical differences in . . . the mental capacities of different human groups." Sir Ronald Fisher, a founding father of modern statistics (and mentor of Diversity Project leader Luca Cavalli-Sforza), perhaps put the point most strongly when he argued that "human groups differ profoundly in their innate capacity for intellectual and emotional development," and this presents the world with the "practical international problem [of] learning to share the resources of this planet amicably with persons of materially different nature" (UNESCO 1952a, 26, 27).[42]

The second significant change to the 1951 Statement was to remove the claim that biological studies support an ethic of universal human brotherhood:

> We now have to consider the bearing of these statements on the problem of human equality. We wish to emphasize that equality of opportunity and equality in law in no way depend, as ethical principles, upon the assertion that human beings are in fact equal in endowment. (UNESCO 1952a, 14)

There were, they argued, two problems with the First Statement's attempt to link equality and biology. First, the link was scientifically unsound. Not only did it unjustifiably limit scientific research, but what, some asked, would happen if someone proved that psychological and mental traits actually were hereditary? Then, in the words of physical anthropologist Carleton Coon, "you are out on a limb." Second, the claim presumed the wrong measure of human equality. Arguments for the fair treatment of races should not be based on what Muller called the "spurious notion" that they are identical in the genetic basis of their psychological and genetic traits. Some even went so far as to liken this attempt to the National Socialists' "notorious attempts to establish certain doctrines as the only correct conclusions to be drawn from research on race, and their suppression of any contrary opinion" (UNESCO 1952a, 57, 54, 32).

As these two revisions to the 1950 UNESCO Statement demonstrate, most physical anthropologists and geneticists believed that the way to construct a credible natural and social order in the wake of eugenics and the Holocaust was not by arguing that biological concepts of race were meaningless for understanding traits that society found meaningful (i.e., "mental," "character," and "personality" traits). Rather, it was by maintaining a boundary between science and society that would enable scientists to define and use race in a manner that could not be used for any particular social end—whether it be totalitarianism or equality. Biological concepts of race, they argued, did not have any inherent fixed social meaning. This is, of course, a very different claim from the claim that race is biologically meaningless, a claim social scientists and historians would later read into the UNESCO Statements on Race.

CHAPTER 2

The Population/Typological Distinction: Physical Anthropologists and Geneticists Contest the Meaning of Race

Differences existed not only between the social scientists who wrote the First UNESCO Statement on Race and the physical anthropologists and geneticists who wrote the Second Statement, but also within each of these two groups UNESCO had deemed experts on race. The differences among the physical anthropologists and geneticists are the most important for an understanding of the Diversity Project debates, as they reveal the wide range of views about race that persisted beyond the purported demise of race in 1950, and shaped the terrain upon which Diversity Project proponents would attempt to build their initiative.

To bring these differences into relief, recall the dominant narrative of the history of race and science: during the first half of the twentieth century a fundamental change in scientific approaches to the study of human diversity took place: the nineteenth-century typological concept of race that cast human differences as static and unchanging was replaced by population and culture, concepts that emphasized the dynamic and changing nature of human differences; scientists who studied human variation (e.g., physical anthropologists and geneticists) shifted their energies away from the construction of mind-dependent hierarchical and typological systems of classification that acted ideologically to support reprehensible social interests (namely, those of slave owners), and instead devoted their energies to experimental and empirical studies that brought to light biological and cultural processes that gave rise to human differences; understanding the dynamic and fluid world of biological and cultural phenomena, and not constructing static typological taxonomies of race, became the order of the day.

Stocking, Stepan, and Barkan all attribute these changes to the eradication of the distorting effects of Nazi and eugenic ideologies from science, and of the influx of the more objective methods of population genetics into the study of the origin and diversity of the human species. By 1950, or shortly thereafter, these analysts argue, scientists had freed studies of human variation from the grip of ideology, and revealed the problems "inherent" in "the concept of biological races" (Silverman 2000, 18).[43] The old nineteenth-century "typological view" of human differences had been cast off and replaced by "the new physical anthropology of adaptive and evolutionary processes" (Goodman and Hammonds 2000, 37).[44] The " 'old' racial anthropology" met its demise, and "a new, non-racial, population genetical science of human diversity" took its place (Stepan 1982, 171).

In its broad contours, this account agrees with the account advanced by supporters of what at the time was commonly called the "population approach." Writing in a 1966 edition of the *Annals of the New York Academy of Sciences* entitled "The Biology of Variation," the University of Pennsylvania physical anthropologist Francis E. Johnston observed that the "population approach" had "brought about an entirely different way of viewing nature" and had led to "an almost total rejection of the typological approach" (Johnston 1966, 508). However, this is where agreement ended. Although most (but not all) believed that the population approach had replaced the typological approach, much debate remained as to what exactly constituted this new scientifically rigorous approach, and how it differed from the old ideologically misguided one.

Prominent historians of science explain the difference by referring to a distinction between Platonic essence and Darwinian variation. The typological approach drew upon a natural classification system that dated back to Plato and Aristotle. According to this classification system, organisms in the natural world come packaged in discrete, static groups that increase in complexity as one moves "up" the "great chain of being" (Stepan 1982, 12, 47). This view of nature opposed Darwin's theory of evolution (Hull 1965, Sober 1980, Stepan 1982, 47; Caspari and Wolpoff 1997, 316–17).[45] Darwin posited that evolution resulted from selective forces acting upon natural variation. Variation, he argued, was vital to understanding evolution. However, the typologist considered variation a mere deviance from underlying essences in nature (i.e., types), and thus showed little interest in Darwinian theories.

More specifically, historians of race and science draw upon five dichotomies to distinguish an ideological typological approach from a scientifically sound population approach: race/population; race/culture; classificatory/empirical; history/natural selection; phenotypic/genotypic. Drawing on these dichotomies, they argue that the old typological approach sought to classify morphological (or phenotypic) differences into static races thought to be the outcome of evolutionary history, while the new population approach sought to investigate experimentally the genotypic differences in dynamic populations thought to be the outcome of the interplay between cultural and biological processes (Stepan 1982, Barkan 1992).

Yet despite the seemingly coherent nature of these approaches, and their stark differences, a closer look at a key episode in post–World War II struggles over the meaning and utility of a category of race in science reveals less clarity. Below I draw upon the debates triggered by the physical anthropologist Frank Livingstone's proposal to adopt a so-called clinal approach to the study of human diversity to demonstrate that each point of purported distinction between the population and typological approach generated debate. The fault-line again formed around disagreements over how to draw the line between ideological approaches to the study of human diversity and rigorous scientific ones.

POPULATION: NEW WINE IN OLD BOTTLES?

"There are no races, there are only clines," Frank Livingstone wrote in the pages of *Current Anthropology* in 1962 (Livingstone 1962, 279). For some, his words marked a change in how the field of anthropology understood and studied human variation (Marks 1996). On one historical account, anthropologists had conventionally treated human variation as if it came packaged in discrete units called races. Livingstone refuted this conception of discrete human races by demonstrating the way in which human variation spills across population lines, forming gradients that Livingstone called *clines*. He suggested that anthropologists should plot the frequency of individual genes—as temperature is plotted on a weather map—noting gradients, and how these gradients correspond to local environmental variation. Based on this clinal analysis, he argued that anthropologists should give up the project of classifying populations into races. Livingstone argued that if one used this approach, "the concept of race"—a concept central to the anthropological study of human beings for nearly two centuries—would no longer be needed (Livingstone, 1962, 279).

Theodosius Dobzhansky, a founder of the field of population genetics and a noted anti-racist, responded to Livingstone's radical thesis in the same issue of *Current Anthropology*:

> I agree with Dr. Livingstone that if races have to be "discrete units," then there are no races, and if "race" is used as an "explanation" of the human variability, rather than vice versa, then that explanation is invalid. Races are genetically open systems while species are closed ones; therefore races can be discrete only under some exceptional circumstances. (Dobzhansky 1962, 279)

In this passage, Dobzhansky took issue with one of the central assumptions underlying Livingstone's call for the end of the use of the concept of race in science: the scientific concept of race packaged human variation into discrete units. Dobzhansky offered another concept of race—a concept grounded in the genetics of natural populations, which treated race as an open genetic system in which races overlapped and changed over time. This population approach to the study of race held that racial differences were not discrete and qualitative (as Livingstone had described them) but rather overlapping and quantitative. In other words, they were not defined by traits that the rest of humanity did not have. Rather, races were populations that contained the same traits. Their point of distinction was the frequencies with which these traits appeared.[46] Thus, Dobzhanksy defined a race as a group of interbreeding individuals (i.e., a population) that differed in the "frequencies of one or more, usually several to many, genetic variables" (Dobzhansky 1962, 279).

This exchange between Dobzhansky and Livingstone calls attention to a defining element of the debates about race and human variation that would later shape the Diversity Project debates: there is no such thing as the concept of race in biology; rather, there are multiple biological concepts of race. Not all believed that race could be used only to order static things, and thus could only produce typological analyses, analyses of no use for studying dynamic biological populations. One could be a populationist (indeed, one could be a founding father of population genetics) and still find race useful, as Dobzhansky did. As he stated one year later in his review of the physical anthropologist Carleton Coon's treatise on race, *The Origin of Races*: "Most biological species are composed of races, and *Homo sapiens* is no exception" (Dobzhansky 1963, 169–70).[47] Dobzhansky did not disagree with Coon's assertion that races exist, or even his belief in their importance for human evolution. The relevant question for Dobzhanksy was not whether race was a valid scientific concept, but rather which concept of race was valid. He advocated for a populationist concept in which race did not refer to a discrete type, but rather to a natural population.[48]

Dobzhansky reserved use of the term "typological" to characterize theories that in his view misused race (Gannett 1999). He then distinguished these misguided typological approaches from the population approach. The latter acted to counter racism by calling attention to the fluidity and internal variability of groups.[49]

This construction of the population approach as anti-racist would gain notoriety in the 1970s when the population geneticist Richard Lewontin wrote an article on the apportionment of human diversity. Lewontin demonstrated statistically that within-group differences in the human species were greater than between-group differences (Lewontin 1972). This article would in subsequent decades be widely cited by scientists and nonscientists alike as an important piece of work that could be used to counter racist claims in society.[50]

At the time of the emergence of the population approach, however, it was not at all clear to scientists that population genetics represented a whole new way of looking at human variation that would lead to the decline of typology and scientific racism. Again, the debates over the clinal approach make this point evident. Loring C. Brace, for example, who along with Livingstone advocated for the clinal approach to the study of human variation, argued that even when race was "conceived in terms of the breeding population," it was the "last bastion of the 'type concept' " (Brace 1964, 314). Like race, population, he argued, acted to obscure patterns of natural variation, and thus impeded the study of the mechanics of evolution.

This uncertainty about whether the concept of population escaped the problems of typological notions of race would not be resolved by Brace, Livingstone, Dobzhansky, or the other physical anthropologists and geneticists who debated race in the early and mid-1960s. Indeed, the debate would continue through the coming decades and resurface at the very first planning meeting of the Diversity Project.

CULTURE: POLITICAL CORRECTNESS OR SCIENTIFIC ADVANCE?

A similar set of ambiguities surrounded the concept of culture. The most noted proponent of the culturalist approach was another widely recognized anti-racist, the physical anthropologist Ashley Montagu. In 1942 Montagu published *Man's Most Dangerous Myth: The Fallacy of Race.* In this book he argued that scientists needed to move away from the "reductionist malaise of regarding populations of human beings as biological races from the purely zoological point of view," and toward an understanding of human populations as the amalgam of biology *and* culture (Montagu 1942, 317). To facilitate this change, Montagu advocated for the replacement of the term 'race' with the term 'ethnic group.' To be clear, Montagu did not argue that human races did not exist. Instead he believed that races needed to be re-conceived as an amalgam of biology *and* culture; this change in conceptualization required a new term.[51]

Loring Brace, however, disagreed with Montagu, arguing that Montagu's use of the term ethnic group was indistinguishable from his use of 'race.' Brace believed that the term had not been introduced to refer to a new concept in science, but only to address the problem of "social injustices which have been perpetrated" in the name of race (Brace 1964, 313). In other words, the change did not result from a change in scientific ideas or practices, but rather social and political pressures.

The geneticist Leslie Clarence Dunn—at the time a colleague of Dobzhansky's at the Institute for the Study of Human Variation at Columbia University—agreed with Brace's allegation that many had abandoned the use of the term race for political reasons, and not because anything in science had warranted the change. As he explained: "Race, in popular usage, is a word with many shades of meaning and connotations and has become so emotionally loaded that some scientists would like to do away with it altogether in referring to human groups." However, unlike Brace,

Dunn believed the term continued to serve a clear and important role in evolutionary biology. As Dunn continues:

> But to the evolutionary biologist it has a clear and unambiguous meaning. I shall use it in this sense: a race is a population which differs from other populations in the frequency of some of its genes (a population has already been defined, for crossbreeding species, as a community of genes shared by interbreeding within the group). The biological concept of race is, thus, a flexible and relative rather than a fixed and absolute one, and this is required by the function it serves in evolutionary thinking. Its function is to identify a stage or stages in the evolutionary process. (Dunn 1959, 90–91)[52]

Like Dobzhansky, Dunn argued that race was a concept "required" by the evolutionary biologist to make sense of "a stage or stages in the evolutionary process" at the subspecies level. By no means did Dunn believe that a study of populations had replaced race; rather, population studies were used to study race formation. Indeed, as a professor at the Institute for the Study of Human Variation, Dunn attempted to understand "changes in races or natural varieties of man" (which he called "microevolution") through the study of isolated human populations (which he labeled "isolates") (Dunn 1959, 114).[53] This research very closely resembled the research envisioned by the first proposers of the Diversity Project.

As these exchanges make clear, physical anthropologists and geneticists who studied human variability during the 1950s and 1960s did not view the use of race as an analytic tool a sufficient reason to label a scientist's work typological. Indeed, many who used what they called a population approach—including those credited with founding this approach, like Dobzhansky—believed that race was the appropriate conceptual tool with which to talk about and understand evolution at the subspecies level. Conversely, employing the populationist's concept of population to characterize human variation did not automatically protect one from being labeled a typologist. Clinalists like Livingstone and Brace, for example, accused populationists like Dobzhansky of continuing the typological legacy of race.

In short, the distinction between the old typological approach and the new population approach does not provide a stable foundation for assessing the history of race and science. As the philosopher and historian of biology Lisa Gannett has argued, this distinction "cannot simply be regarded as a conceptual tool for historical analysis but is itself a product of history whereby these two modes of thought came to be categorized and opposed to one another." Rather than switching from a typological concept of race to the population approach to the study of human diversity, the typological-population distinction "came to be employed instrumentally" by scientists who in the 1950s and 1960s sought to distinguish their use of the concept of race from the legacy of the abuse of race by the Nazis in Europe and by segregationists and eugenicists in the United States (Gannett 1999, 2).

CLASSIFICATION: TOOL OR IDEOLOGY?

Other dichotomies used to distinguish the typological from a population approach also shift precariously on closer inspection—for example, classifying variation versus empirically investigating it. Most prominent historians of these ideas have argued that the typological approach was part of "the old, *classificatory* biology of race" that was replaced by the "new, *evolutionary* biology of Man" (Stepan 1982, 174; emphasis added). The problem with the old biology of race was that it held out classification as a primary goal. Race declined when it became clear that there was no consistent or objective method for classifying human variation, and scientists realized that classification was "hopelessly entangled with themes of ethnicity and nationality" and thus an inappropriate goal for the scientist (Barkan 1992, 3). The "new science of human populations" opposed this classificatory biology by casting off analyses based on ideal types and replacing them with analyses based on empirical studies of nature (Stepan 1982, 173, 181).

This understanding of classification as a social project that got caught up in the nettlesome thicket of ethnicity and nationality, and not as a scientific endeavor, echoes the views of clinalists like Livingstone. Livingstone argued that human variation defied classification efforts because human traits were discordant. In other words, one human character did not reflect the variability of another character. Thus, any attempt to classify human variation by race would be arbitrary and, even worse, impede explorations of the origins of human variation. Livingstone cited this difficulty of classifying races as a further sign that races did not exist (Livingstone 1962, 279).

Dobzhanksy disagreed with Livingstone's analysis. He did agree that there had been a multiplication of names for races, and that these names indeed did not reflect objective criteria. However, he did not believe that this meant that races did not exist, or that one should give up on classification.[54] In order to understand why, he argued, one had to distinguish between race as biological phenomenon and race as a category of biological classification. One could objectively ascertain the existence of biological races without engaging in the project of classification. The first act—discovering races through empirical study—was a biological problem. The second act—categorizing and labeling the discovered races—was a nomenclatorial problem. The two problems, Dobzhansky argued, should not be confused. That different scientists used different systems for naming races, he explained, should neither be surprising, nor call into question the reality of race. Classification, after all, did not arbitrate reality, he argued. Rather, it served as a tool that scientists used "in writing and speaking" to ease discussions of populations or organisms under investigation (Dobzhansky 1962, 280). Different scientists could come up with different systems of classification, he explained, because they were interested in different aspects of biological phenomena that could be more easily explored using one classification system rather than another. The Polish physical anthropologist Andrzej Wiercinski concurred: "It is not important that a great number of classificatory systems exist due to different sets of traits considered; each system reflects a different aspect of human variability" (Wiercinski 1964, 318).

Dobzhansky believed that population genetics provided a basis for sound systems of classification. If the problem before had been that classification systems based on ideal types did not correspond to nature and obscured biological processes, then, he argued, population genetics provided a solution by placing biological processes first, allowing them to dictate categories of classification. So, for example, he recommended that the naming of races should follow the discovery of a barrier that limited gene exchange between populations—whether it be national, ethnic, religious, linguistic, or class boundaries.[55] Dobzhansky and his colleagues argued that these new biologically based categories would avoid the problem of erroneously equating ideal types with natural types. Nature would determine categories rather than human-constructed categories being imposed on nature. As Dobzhansky's colleague, the geneticist Leslie Clarence Dunn, explained:

> [W]e put things in pigeonholes, in studying evolution at least, in an effort, not to attain the order of tidiness, but to try to understand the pattern of the whole variety. We put together, as members of a race, populations which have many, perhaps most, of their genes in common. We want, not order for its own sake, but order for the sake of tracing relationships and evolutionary descent. So, for our purposes, the use of race as a classifying device is secondary to and determined by its use in helping us to unravel the story of how man attained his present variety. (Dunn 1959, 91)

Yet more than thirty years later, when the Diversity Project was first proposed, scientists still had not resolved how to place nature first, and classification second. As we will see, organizers of the Diversity Project ran into trouble as nature refused to disentangle itself from human-constructed systems of classification. Rather than using studies of nature to direct the creation of classification systems, studies of nature and classification systems would have to be produced together, each influencing the other.

Dobzhansky's defense of categorization served as a resource for other scientists who attempted to respond to the clinalists' critiques of the use of race. For example, in his response to Brace in *Current Anthropology*, the physical anthropologist Earl C. Count argued that Dobzhansky, "an incomparably finer geneticist than are the anthropologists who have seized occasion to take him to task," had provided a "rationale" for " 'racial classification.' " In response to Brace's accusation that taxonomic principles below the species level are either invalid or futile, Count responded: "On what grounds, pray?" (Count 1964, 315).

HISTORICAL CONTINUITY OR CONTINUAL ADAPTATION?

Not only did most populationists believe in categorization—despite the fact that populations overlapped and were in a constant state of flux—they also believed that categorization could and should reflect historical processes. Stepan has argued that attention to historical events was a hallmark of the old typological theory of race. According to this theory, racial differentiation of the human species resulted from a series of events in prehistory, and since that time the human species had undergone little organic change (Stepan 1982, 178). Stepan holds that the new science of human populations challenged this understanding of human variation by demonstrating the ways in which variation arose from a process of continual adaptation to the natural environment.

For scientists at the time, however, the available theories of human variation were not so distinct. Some anthropologists, for example, argued that one did not have to choose between the bio-historical concept of race and a population concept of race. Ernest Hooton's student Stanley Garn, for example, contended that in order to understand "the very problems of natural selection" (the problem of central interest to the population geneticist), one had to consider the "residual store of genes in particular populations" (Garn 1964, 316).[56] In other words, history mattered, especially to scientists interested in the processes that structured the constant flux and change of human variation.

PHENOTYPE AND GENOTYPE

Finally, the distinction between phenotype and genotype provides no stable ground for demarcating a typologist from a populationist. Historians of biological beliefs about race have argued that the old typological racial science defined race "anatomically and morphologically, in terms of the phenotype—that is, by detailed measurements of the shape of the skull, the dimensions of the post-cranial skeleton, by stature, and by skin colour." Populationists, on the other hand, studied populations that were "defined not morphologically or behaviourally but genetically and statistically" (Stepan 1982, 176). This account does match those who supported a population approach to the study of human variation. Dunn, for example, defined physical anthropologists as those who studied phenotype, while geneticists studied genotype (Dunn 1959, 117). Additionally, Montagu distinguished between race defined "in terms of absolute phenotypic differences" and race defined in terms of "relative differences in the frequency of genes"—attributing the former definition to old anthropologists stuck in a "pregenetical phase" of biological research, and the latter definition to geneticists and the new physical anthropologists (Montagu 1950, 316–22).[57]

Yet many physical anthropologists associated with the "old" school did value genetics. Indeed, many were eager to do genetic research. However, they did not believe that they should replace conventional anthropological methods. As Wiercinski explained: "We in Poland are familiar enough with classic as well as modern genetic literature. But we simply do not regard some primitive statistical models of population genetics as untouchable idols" (Wiercinski 319, 1964). Wiercinski and other physical anthropologists rejected the idea that the only valid contemporary studies of human variation and race were genetic ones. To study only genetic traits, they argued, was unrealistic, since the genetics of too few human traits had been worked out to make genetic analysis feasible. It was also undesirable, since even if the genetic basis of all traits were known, genetic analysis still would not allow for the analysis of the "human organism as a whole" (Hoebe quoted in Montagu, 1950, 323; Coon quoted in Silverman 2000, 18).[58] For these reasons, they argued that morphological analyses would continue to provide invaluable information.

Many physical anthropologists also rejected the idea that genotypic analyses escaped the problems of typology associated with phenotypic analyses. As Rachel Silverman pointed out in her study of post–World War II blood group analysis, William Boyd, one of the innovators of genotypic analyses of blood types, chose the populations he sampled based on morphological (or phenotypic) criteria. This led some to question if purported genetic classifications of races could be deemed any more "objective" than morphological classifications (Silverman 2000, 14).

Much of what was at stake in these debates was a disciplinary turf war over who had the right tools and concepts for studying human variation. Traditionally human variation and evolution had been studied by physical anthropologists through analyses of the morphology of human fossils (or phenotypic differences) (Barkan 1992, 4). During the 1920s and 1930s, however, the new science of population genetics emerged.[59] Physical anthropologists such as Coon, Wiercinski, Hooton, and others worried that the rise in popularity of genetic studies of human variation might lead to the takeover of physical anthropology by geneticists. They believed this takeover would result not from the superior knowledge of geneticists (to the contrary, many considered geneticists dilettantes when it came to the study of human variation), but rather from the changes in the political currents following World War II (Silverman 2000, 19). Argued Coon in a speech given at Brown University in 1965:

> Until well into the 1930s, most physical anthropologists were content to measure scores of human populations. . . . Then they discovered genetics, largely through the medium of blood groups and it became fashionable to downgrade such troublesome polygenic products as metrical and morphological constants and put all their eggs in the monogenic basket of hematology. This shift was particularly opportune because of the rising tide of prejudice against racial prejudice. *Blood groups were politically and socially safe and respectable* [emphasis in original]. (Coon quoted in Silverman 2000, 18–19)

As this excerpt makes clear, Coon viewed anthropologists as the victims of what would today be called political correctness. The problem was not that anthropologists were living in the backwaters of "pregenetical" science—as Montagu and others suggested—but rather that political changes had made the good work of morphological and metrical analysis politically sensitive (Montagu 1950, 318; Wiercinski 1964, 319).[60] Forty years later, at the time of the Diversity Project's proposal, these tensions had not been resolved.

As all of these debates between physical anthropologists and geneticists demonstrate, no consensus about the role of race in studying human origins and diversity emerged following World War II.[61] Physical anthropologists and geneticists did not all agree—contrary to prevalent historical opinion—that race had no biological meaning, and should be replaced by a study of populations.[62] Not even did all agree that typologies had no use in science. Rather, most sought to redefine scientific ideas and practices for studying race (including typologies) in the wake of what many perceived as the abuse of these ideas and practices by eugenicists, segregationists, and the Nazis.

Conclusion

Segregation and race-based lynchings in the American South, eugenics movements, and above all World War II and the Nazi Holocaust sparked fundamental questions about the meaning and place of race in society in the second half of the twentieth century. Historians, however, have paid less attention to the fundamental questions that these events stirred about the scientific meaning and foundations of race—questions that would continue to be of importance in biology and anthropology well beyond the purported demise of the category of race. The critiques of race that began in the interwar period did not hasten its demise, as some observers of the history of biology and anthropology have argued. Instead, it led scientists to revisit fundamental questions raised by its use as an ordering device in human diversity research: What is the proper role and meaning of systems for classifying humans? How should human differences be defined and for what purposes? To what extent do these differences reflect unchanging biological or social characteristics, and to what extent are they a part of a dynamic system of cultural and biological change? When, and in what respects, do systems of classification artificially and harmfully represent human differences, and when, by contrast, do they provide useful tools for understanding those human differences? As this chapter has shown, these questions proved just as problematic for scientists as for other actors in society.

Ongoing struggles around these fundamental questions did not lead most scientists to conclude that race should be "delegated to the scientific scrap heap with the advent of studies of evolutionary and adaptive processes" (Goodman and Hammonds 2000, 29). Instead, it led many to redefine race in the wake of scientific and political developments. Both physical anthropologists and geneticists would continue to believe that race served an important function in the study of human variation and evolution. Dunn, for example, argued that race functioned to identify a stage or stages in the evolutionary process. His view was widely shared.

Instead of rejecting the use of race as an analytic category in science, natural scientists would argue that race needed to be redefined and retooled to reflect the precision of the new science of population genetics. In adopting the ideas and practices of this new science they did not displace race. Rather, they crafted a new concept of race defined in populationist terms. Instead of referring to static essences, population geneticists and the "new" physical anthropologists argued race should be used to refer to subgroups within a dynamic system of populations.

Just how new this population concept of race was remained a source of debate among physical anthropologists and geneticists. Not all believed that the population approach to race escaped the problems of the old typological anthropology. Further, not all agreed that the so-called old anthropology should be abandoned. Indeed, many believed that analyses of phenotypic differences, historical studies, and systems of categorization (even ones that were typological) should still play important roles in the study of human variation and evolution. Given these hesitations on the part of population geneticists and physical anthropologists about abandoning older modes of analysis, to argue that population studies led to a "paradigm shift" or a "fundamental change" or a "revolution" in biology perhaps overstates the case (Goodman and Hammonds 2000, 29; Stepan 1982, 176; Barkan 1992, 342). Even with these new tools, methods, and concepts, debates about the role of race in science—and attendant questions about how and when to classify human differences—would continue. Although some of the practices and ideas for studying race and human diversity would change, and new concepts of race would emerge, many of the fundamental questions and issues would endure.

These questions and issues would not be resolved by the time the Diversity Project was proposed forty years later. Indeed, to understand the Project's proposal, we have to situate it in the context of these earlier efforts to study and order human diversity. It is to this task that I now turn.

Chapter 3
In the Legacy of Darwin

As noted in the last chapter, the rise of the category of population in biological studies of human variation did not mark the demise of the category of race in biology. Indeed, for many of its advocates, population genetics represented a source of new, more rigorous tools for studying race formation. In this chapter, I provide the historical context that renders visible the connections between population genetics, studies of race formation, and the Human Genome Diversity Project.

At the time of its proposal, the ties between the Diversity Project and decades of previous research on the formation of human races in biology and anthropology were anything but transparent. The original call for the initiative published in the journal *Genomics* made no mention of human races or the desire to study them (Cavalli-Sforza et al., 1991, 95). In the political and institutional environment of the time—an environment marked by the end of the Cold War and the emergence of what some scholars have named a "market for humanitarianism"—human similarities, not racial and ethnic differences, gained political currency (Rabinow and Palsson, 2001).[63] World leaders, such as the then U.S. president George Bush, called for global unification and a "new world order." Molecular geneticists called for an international effort to sequence "the human genome." For their part, the population geneticists who proposed the Diversity Project called for an understanding of "our species," and argued that studies of human genetic variation would support humanitarian and anti-racist causes (Cavalli-Sforza et al., 1991, 490).

Indeed, in the fall of 1989, just a year before efforts to organize the Diversity Project began, Luca Cavalli-Sforza wrote a letter to a central organizer of the then nascent Human Genome Project (HGP) explaining that human population genetics research served to undermine notions of race that fueled racism and discrimination in society.[64] Cavalli-Sforza's letter responded to fears expressed by leaders of the Genome Project that their initiative might create new anxieties about eugenics and racism. Concerns about a potential new racism sparked by human genomics, and assurances by population geneticists such as Cavalli-Sforza that they were not warranted, formed an important part of the backdrop for both the Human Genome Project and the first call for the Diversity Project.

Although recognized by members of the genomics community for critiquing what they labeled social concepts of race, Project proponents' uses of what they understood to be genetic concepts of race passed with little or no notice. The population geneticists who first proposed the Project employed these latter concepts as they wrote grant proposals and journal articles that they hoped would generate scientific support for the initiative. At the time, these concepts of race generated no special attention.

To explain these different concepts of race, and their uses by Diversity Project organizers, this chapter situates the initiative within the broader context of the history of the Project's original disciplinary home, human population genetics. The field of human population genetics arose to study and explain the nature and source of inherited differences among individuals within a species (Cavalli-Sforza and Bodmer 1971 [1999], xi). As we learned in the last chapter, scientists in this field commonly used the term race to denote subspecies categories in which these differences evolved. The relationship between the Diversity Project and previous attempts by human population geneticists to study the formation of these subspecies groups, or races, is the primary topic of this chapter.

In particular, I will examine three central texts in biology: the report of the 1950 Cold Spring Harbor Symposium (CSH), *Origin and Evolution of Man*; the 1971 edition of *The Genetics of Human Populations* (written by Luca Cavalli-Sforza and his colleague, Walter Bodmer); the 1999 reprinted edition of *The Genetics of Human Populations.* As these texts illustrate, the category of race played a central role in organizing genetic analyses of human history and evolution. Conceptually, these texts closely resembled those of the original proposals for the Diversity Project. Indeed, I argue that the proposal for the Diversity Project in 1991 and the proposals to study the formation of human races in 1950 and 1971 are all part of the same long-standing effort within biology to use genetic techniques to understand evolution within the human species.

The Human Genome Project and the Proposal for the Human Genome Diversity Project

In October 1990 the U.S. National Institutes of Health and Department of Energy authorized $3 billion to be spent over fifteen years on a project to sequence the human genome.[65] Although the resulting map of sequence might look like a dull scroll of four letters, in the June 1990 issue of the *Journal of the American Medical Association* (*JAMA*), James Watson, co-discoverer of the double helix and the director of the newly formed Human Genome Project, and Robert Cook-Deegan, his policy adviser, promised nothing less from the map than a "new powerful foundation of knowledge" upon which biomedical researchers could launch "their assault on disease." The sequence of the human genome, Watson, Cook-Deegan, and many other proponents of the Human Genome Project argued, would lead to the development of molecular tools needed to cure diseases that "exact an enormous toll of human suffering on every culture and in every geographic region." It would help all human beings, and thus would be a " 'public good' in the best sense" (Cook-Deegan and Watson 1990, 3322).[66]

Further, Watson and Cook-Deegan argued, the Genome Project would not—as some early critics charged—be an American project that served only American interests (Interview N). It would not discriminate on the basis of national or cultural boundaries. Instead, its benefits and organization would be "inherently international." Indeed, it would necessitate a "coordinated worldwide effort to share resources, to spread the burden of funding the research, and to take advantage of unique resources that can be found anywhere in the world" (ibid., 3324).

Leaders of the Human Genome Project were not alone in looking to the promise of global initiatives in 1990. Only a year had passed since the fall of the Berlin Wall. Optimism about unification and a "new world order" filled the speeches of leaders in the United States and Europe. Fittingly, at a time when world leaders were reaching across the divisions of the Cold War, geneticists first proposed an initiative to sample human genomes from around the globe.

At first, the proposal took up just one paragraph in Watson and Cook-Deegan's June 1990 *JAMA* article:

> Some populations in the Middle East, Asia, and Africa have been isolated by geography and limited technology, and represent rare and valuable resources to study human origins and patterns of population genetics. We must take steps, through international collaborations, to ensure that DNA is collected and stored for subsequent analysis, before populations are lost for all time by the ingress of technology, the press of

> population growth, and other factors that break down the traditional barriers that have isolated these groups. (Cook-Deegan and Watson, 1990, 3323)

This passage reflected the concerns and themes of a talk given by Luca Cavalli-Sforza that Cook-Deegan had heard in 1989 as he attended scientific meetings trying to organize the Genome Project. In that talk, Cavalli-Sforza described his hope to organize an initiative to collect and store DNA from "vanishing" indigenous populations whose genomes he believed contained the answers to questions about human evolution and origins.[67]

The Project resonated with Cook-Deegan for a number of reasons. First, he believed Cavalli-Sforza was right: the populations that Cavalli-Sforza was interested in sampling *were* vanishing. This he assessed from firsthand experience working with some of these populations—in particular the Kurds. Cook-Deegan had gone to Turkey in October 1988 to help the Kurds after the end of the Iran-Iraq War.[68] Saddam Hussein had been systematically killing off Kurdish males (the Kurds had assisted the Iranians against him in the Iran-Iraq War), destroying villages, and moving populations. There were initial reports about the use of poison gas, but then a second wave of press accounts cast doubt on those stories. In the fall of 1988, Cook-Deegan was a member of a three-person medical team that interviewed refugees in camps in southeast Turkey and then wrote a report entitled "Winds of Death" for Physicians for Human Rights.[69] It was in Turkey that Cook-Deegan learned about the threat to the future survival of the Kurds.

Second, as a doctor trained in genetics, Cook-Deegan believed that populations like the Kurds could be important to medical science. As he and Watson explained in their *JAMA* article, the "cultural practice" of consanguineous marriage made populations in the Middle East, Asia, and Africa of interest not only to population geneticists, but also to medical researchers: "It is likely that many hitherto unrecognized genetic disorders reside in such populations, because the expression of recessive mutations is much more likely with intermarriage" (Cook-Deegan and Watson 1990, 3323). Expression of these mutations would then make disease genes easier to find.[70]

Finally, 1990 was a crucial year for genome budgets and the Genome Project had come under scrutiny for appearing to be primarily geared to American interests. Cook-Deegan believed that the sampling project, with its international scope, might help to counter this critique. It might also increase the Genome Project's appeal to biologists, many of whom had argued from the beginning that mapping and sequencing the human genome would not be valuable in and of itself (Cook-Deegan 1994).[71]

For all of these reasons, just a few months after the publication of the *JAMA* article, Cook-Deegan agreed to write the first draft of a call for a worldwide survey of human genetic diversity. The second human genome meeting held in Valencia, Spain, in October 1990 provided the occasion for this initial organizational effort.

A Proposal to Sample Human Genetic Diversity

Financed by a bank in Spain, the second international genome meeting in Valencia was a lavish affair. Hundreds of thousands of dollars were spent to bring thirty to forty scientists together to showcase their research and to build international support for the genome initiative (Interview N; Cook-Deegan 1994). Although the meeting in Valencia was primarily intended to build support for the Human Genome Project, it would also mark the beginning of efforts to organize the Diversity Project.

Among the featured speakers at the meeting was Mary-Claire King, a noted population and medical geneticist whose research on the relatedness of humans and chimps graced the front cover of *Science* magazine in April 1975 (King and Wilson 1975).[72] In Valencia, however, King did not talk about her human evolution research, but about her human rights work with the Grandmothers of the Plaza de Mayo whose grandchildren had been abducted during Argentina's Dirty War in the 1970s. During this war, the Argentinean military kidnapped, tortured, and murdered fellow citizens suspected of " 'unpatriotic subversion' " (Arditti 1999, 14). As part of this campaign of terror, "disappearances" began. Those suspected of political opposition were imprisoned, given a letter, and assigned a number in sequence (e.g., M1, M2). If they used their names, they were tortured. The goal was to strip them, brutally, of their identities. Many died.

In 1977 mothers of the disappeared began a campaign to end the reign of silence that surrounded these brutalities. Every Thursday at 3:30 in the afternoon they went to the Plaza de Mayo in Buenos Aires to march and to ask to talk to the president about the disappearances (Arditti 1999, 35). They also demanded help in finding their grandchildren, many of whom had been kidnapped and placed in purportedly patriotic families at the time of the disappearances. By 1980 the Grandmothers of the Plaza de Mayo had gained considerable international human rights support, but they still faced a significant problem. Several years had passed since the kidnappings of their grandchildren. Photos were outdated. Judges demanded more definitive proof of identity. In an effort to respond to these demands, the Grandmothers met with Eric Stover, then director of the science and human rights program of the American Association for the Advancement of Science (AAAS). Stover consulted with a Chilean scientist working at the NIH at the time, who in turn contacted Mary-Claire King. King developed an array of highly specific genetic markers for proving family relatedness. Using these markers, the Grandmothers won at least fifty court cases (Bass 1994, 220).[73]

At Valencia, King's account of her work with the Grandmothers moved to tears the distinguished members of the audience, including the queen of Spain. As one attendee later recalled: "There was not a dry eye in the house when she was done" (Interview N). The speech presented population genetics to the assembled world leaders as a science of liberation. At this moment, it could not have seemed further from the science of repression that critics would later label it.[74]

Organizers of the meeting had intended King's talk to demonstrate the relevance of human genomics—and, ultimately, the Human Genome Project—to humanistic causes. The talk, however, would also provide an occasion for Cook-Deegan to approach King about Cavalli-Sforza's idea of sampling "vanishing" indigenous populations.[75] King, eager to create the human population genetics resource that Cook-Deegan described, encouraged him to write up a draft proposal for what would later become known as the Human Genome Diversity Project. Cook-Deegan acted on her suggestion and called Cavalli-Sforza in Palo Alto, California, to ask him to fax some background materials to Valencia. Cavalli-Sforza sent the final chapter of his draft copy of the *History and Geography of the Human Genes*, a massive volume ultimately published in 1994 that brought together in one place all the data collected in human population genetics (Cavalli-Sforza et al., 1994). On the basis of this chapter, and his prior conversations with Cavalli-Sforza, Cook-Deegan drafted, one evening in Valencia, a very preliminary call for the initiative. He gave the draft to King to revise for scientific accuracy on her plane ride back to San Francisco. King then sent the revised draft to Cavalli-Sforza and the UC Berkeley biochemist and molecular evolutionary biologist Allan C. Wilson—the two scientists Cook-Deegan and King believed would have to support the initiative if it were to succeed.

On 22 February 1991, Mary-Claire King sent out for comment a nearly final draft to Luca Cavalli-Sforza, Allan Wilson, Cook-Deegan, Charles Cantor (principal scientist of the Department of Energy's Human Genome Project), Jared Diamond (UCLA Department of Physiology, author of *The Third Chimpanzee*), and Bruce Ames (UC Berkeley geneticist, famous for his work on the mutagenecity of natural and human-made chemicals). As Project records indicate, in a cover memo headed "Sampling the World," King wrote that she had penned the note on Saturday at noon, and that she hoped that by the time all had received it, there still would be a world to sample. The Gulf War had begun just a few weeks before.

As this memo and other countless acts indicated, humanitarian concerns clearly motivated King and Cook-Deegan. Neither of them was recommending that scientists collect tissue samples from indigenous groups only for preservation before these groups died out, and without regard for their ultimate survival. To the contrary, both of them were exemplary scientists who in their work with Amnesty International, Physicians for Human Rights, and the Grandmothers of the Plaza de Mayo had demonstrated their dedication to helping oppressed peoples of the world. Indeed, it would be hard to imagine two scientists who better exemplified a "model citizen-scientist."[76] To support the proposed sampling initiative, both of them must have been convinced of its political as well as its scientific merits.

This is not to say that Cook-Deegan, at least, did not worry that the project might raise political and ethical concerns. Indeed, in his first draft of the call for the initiative he included a paragraph on North/South relations, the need for sampled populations to participate in the research, and the possibility that such research might create the potential for stigmatization. Cavalli-Sforza, however, removed most of this paragraph from subsequent drafts and convinced Cook-Deegan that King's work with the Grandmothers in Argentina, and Nancy Wexler's work on Huntington's disease in Venezuela, provided two of many examples that demonstrated that population genetic research would benefit people; there was, he argued, no reason for special concern (Interview N, "Commentary: The Urgent Need" 1991).[77]

Cook-Deegan later worried that the initiative might be viewed as a "sample and run" project, what *Science* magazine reporter Leslie Roberts would describe as "a planeload of Western geneticists descending on the jungle, collecting blood, and then disappearing " (Roberts 1991b, 1617, Cook-Deegan 1991). To avoid this image he suggested linking the Project to humanitarian relief efforts. The idea was that if doctors associated with these efforts could be taught how to collect information on family and ethnic backgrounds, they could collect blood for the initiative. Such a collection plan, he told Cavalli-Sforza, could help to avoid the resentment created by "sample and run" tactics. He went on to note that Cavalli-Sforza should keep in mind that delivery of services to impoverished people would continue to be the first priority of these organizations. Taking samples would be a second priority, and "quite properly so" (Cook-Deegan 1991).[78]

In the *Genomics* article that first announced the initiative, however, all these discussions of how the initiative would integrate its scientific practices with political and ethical practices were omitted.[79] Indeed, the only substantive matters discussed in this article were logistical ones related to sampling procedures. Blood, hair, or human tissue samples, the authors explained, would be collected from "isolated human populations" of "special interest" around the globe. Because these populations lived "far from airports and modern laboratories," collection would require the creation of regional facilities to transform samples into cell lines. These cell lines would be distributed to "facilities such as the Coriell Institute in the United States, the UN-IDO-supported biotechnology research centers in New Delhi and Trieste, and designated facilities in Latin America and the Middle East." DNA harvested from these cell lines would be used to "illuminate variation, selection, population structure, migration, mutation frequency, mechanisms of mutation, and other genetic events of our past" (Cavalli-Sforza et al., 1991, 490). Only the following three sentences addressed ethical and political concerns:

> Among these very informative groups have been many peoples historically vulnerable to exploitation by outsiders. Hence, asking for samples alone, without consideration of a population's needs for medical treatment and other benefits, will inevitably lead to the same sense of exploitation and abandonment experienced by the survivors of Hiroshima and Nagasaki. It will be essential to integrate the study of peoples with response to their related needs.[80]

The authors of the call for the Diversity Project signed onto a document that contained only this briefest mention of questions about racism and the potential of the sampling effort to stigmatize populations—questions that would later move to the forefront of the concerns of Project organizers and their critics.

Further, the *Genomics* article made no mention of the possibility that the Project might raise complex questions about the meaning of race and human differences in either science or society. Below I situate this absence in the broader political and scientific contexts in which the Project was proposed—contexts in which a new discourse of race held that the end of racism depended upon the explicit *non*-recognition of race. I illustrate how this new discourse connected to changes in scientific practices and ideas that made it possible for geneticists to argue against the biological reality of what they labeled social constructions of race, while at the same time employing what they labeled genetic concepts of race for understanding the evolution of the human species and its diversity.

Race: Visible or Invisible?

From the very earliest days of planning the Human Genome Project, organizers were concerned that the Project might raise new fears about race hygiene and eugenics.[81] Most recognized that although it had been fifty years since World War II, race, eugenics, and human genetics remained tightly linked in the minds of many. Further, many at the National Center for Human Genome Research (NCHGR), the institutional home for the Human Genome Project, knew that Richard Herrnstein and Charles Murray were completing *The Bell Curve*, a book that would revive claims that there was a correlation between race and IQ and would rekindle concerns about scientific racism.[82] It is in the context of these new concerns that Cook-Deegan, then special section editor of *Genomics*, wrote to Luca Cavalli-Sforza in the fall of 1989 to ask him to write a review on what genetic research revealed about the reality of race, and if human genomics could lead to a new racism.[83] Cook-Deegan sought out Cavalli-Sforza—arguably the most prominent living human population geneticist—to write the review because he believed that Cavalli-Sforza "would know the science most relevant to the question [of race]" (Interview with Cook-Deegan 1999).[84]

Cavalli-Sforza responded with a long and detailed letter. In it, he laid out a view of race that resonated in many respects with the one that population geneticists and physical anthropologists had presented in the Second UNESCO Statement on Race in 1951. First, reminiscent of claims made in the First and Second UNESCO Statements, as well as Livingstone's essay "On the Non-Existence of Human Races," Cavalli-Sforza asserted that "no criterion" existed for choosing among the many possible ways of partitioning humans into races. Statistical error, he explained, as well as many other sources of uncertainties, plagued every attempted classification of races. As in the UNESCO Statements, Cavalli-Sforza argued that biological groups overlapped, and that no single trait could distinguish one group from another. So, he concluded, "why classify races if the result is arbitrary and uncertain?" (Cavalli-Sforza 1989).

Additionally, like the Second UNESCO Statement on Race, Cavalli-Sforza argued that genetics could not exacerbate racism for the simple reason that as a "science" it was separate from the "moral" determinations made in "society" (ibid.). Yet although Cavalli-Sforza argued that genetics did not support racism, he stopped short of suggesting that there was no place for a concept of race in genetics. As Theodosius Dobzhansky had explained to Frank Livingstone thirty years earlier, to claim that there are no objective criteria for classifying human races is not necessarily to argue that races do not exist, or that concepts of race are meaningless. Cavalli-Sforza held the same view. He did not argue in his letter to Cook-Deegan that genetics undermined all concepts of race; rather he argued that genetic research undermined "popular" and "social" constructions of race used by what Cavalli-Sforza called the "man on the street." The "man on the street," so Cavalli-Sforza maintained, based his diagnosis of human differences on visible physical traits: for example, skin, hair, and eye color. Yet human population genetic research demonstrated that these visible traits "give a different evolutionary result from that of the analysis of genes" (ibid.). The reason: these traits resulted from relatively recent adaptations of humans to different climates. They reflected environmental changes that hid the vast genetic similarities that lay, literally, underneath the skin, in a realm invisible to the lay observer.[85]

In this way, Cavalli-Sforza's claims about race differed from claims about race made by the physical anthropologists and geneticists who drafted the Second UNESCO Statement on Race. The UNESCO "experts" had held the "man on the street" to be a witness of biological truths, who could, even with his "untrained eye," see that races differed from one another (UNESCO 1952b, 7). In Cavalli-Sforza's letter, however, the "man on the street" figured as an unreliable witness who could not see what the "the scientist" could see, and so was easily deluded into believing erroneous racist ideologies (Cavalli-Sforza 1989). Cavalli-Sforza also departed from the Second UNESCO Statement on Race in his critique of the value of visible physical markers in biological analyses of race. The authors of the Statement had argued to the contrary that a biological notion of race could be defined using visible traits such as hair texture and eye color.

The significance of the differences between Cavalli-Sforza's views about race in 1989 and those of his intellectual predecessors forty years earlier is profound. As is well known to any student of biology, classical Mendelian genetics originally depended upon observing differences in visible physical traits. Mendel, for example, compared wrinkled and smooth peas. He held that these traits were the overt signs of an underlying reality in nature. In 1950 all geneticists still subscribed to the notion that visible physical traits corresponded to underlying genetic characteristics. They called these visible physical traits the phenotype. The phenotype was held to be the realization of the genotype in the observable world. However, as molecular techniques and molecular biology developed in the 1960s, the meaning of the visible and the physical for the geneticist simultaneously changed. Using gel electrophoresis, geneticists could now observe not just hair texture, nose structure, and eye color on the surface of the body, but also molecules within the body's cellular structures. These molecules were held to be closer to the action of the genes, and thus better markers of their biological reality.[86] Indeed, by the 1980s technologies had been developed that rendered visible what had become known as the very physical stuff of genes: deoxyribose nucleic acid, or DNA.[87]

With the advent of these newly discovered molecular markers it became possible for biologists—such as Cavalli-Sforza—to relegate physical traits visible to the "man on the street" to the realm of unscientific analyses of race and human diversity. In other words, reversing earlier presumptions about a one-to-one correspondence between phenotype and genotype, biologists were now prepared to argue that surface differences hid deeper similarities. Thus, the lay observer might believe based on skin color and hair type that human beings are split into fundamentally different biological groups—namely, races. However, population geneticists argued that the genes revealed a different story: for many hundreds of thousands of years—before they migrated out of Africa—human beings had lived together on the same continent, under the same sun. During that time members of the species essentially evolved together as one family; the similarity of human genotypes reflects this common history (Cavalli-Sforza 1989).

This reworking of meaning made it no longer necessary to make a distinction between scientific and unscientific interpretations of *visible* physical traits, as the UNESCO "experts" had done. With the advent of molecular biology, geneticists could argue that they no longer needed to rely on visible physical traits (like hair color) to study evolution.[88] Just as the importation of statistical techniques had made it possible for scientists to argue that they had tossed typological notions associated with racism into the dustbin of pseudoscience, so the rise of molecular techniques made it possible to do the same for concepts of race propagated on the basis of visible physical characteristics.

These changes in the scientific construction of race must be understood in their broader political context. As the American historian Peggy Pascoe has argued, by the 1960s competing ideologies of race had been "winnowed down to the single, powerfully persuasive belief that the eradication of racism depends on the deliberate nonrecognition of race" (Pascoe 1996, 48). The 1954 decision of the U.S. Supreme Court in *Brown v. Board of Education* provides perhaps the most famous example of this cognitive shift. In this historic case, the Court unanimously held that "separate but equal" schooling was unconstitutional, and that race should not play a role in determining access to education.[89] Pascoe has given the name modernist racial ideology to the belief in *Brown* and in U.S. society more broadly that race relations would improve if society did not recognize the category as valid. According to this ideology, race is merely a set of physical traits (that include skin color, hair texture, etc.) and should have no meaning in society. The notion that society should grant no significance or meaning to race would later gain prominence in the 1980s during the Reagan-era push for a "color-blind society" (Omi and Winant 1994, 1).[90]

Cultural anthropologists, most notably Franz Boas, also provided intellectual support for this argument about race (Pascoe 1996, 53).[91]

These scientists argued that physical differences mattered little to the development of human traits meaningful in society. Human population geneticists, like Luca Cavalli-Sforza, took their argument one step further. They argued that independent of genetic analysis, physical traits commonly used to distinguish races (e.g., head form and skin color) revealed nothing about human biology and the evolution of the human species. In other words, drawing upon distinctions made possible partially through the advent of molecular biology, population geneticists crafted new discursive resources that enabled them to join lawyers, government officials, and even many social scientists in arguing that race, when it is held to be merely a cluster of visible physical traits, is meaningless—both in science and in society.

This rejection of a visible physical concept of race is *not* inconsistent with the retention of other concepts of race in human population genetics. As demonstrated in the previous chapter, proponents of genetic studies of human populations critiqued some concepts of races while advocating the use of others. Instead of arguing that race did not exist, or that concepts of race had no utility in science, those population geneticists and physical anthropologists most celebrated for their anti-racist goals—like Dobzhansky and Montagu—were often arguing for a reconceptualization of race.

Indeed, as we will see below, in order to win support for their research, Dobzhansky, Montagu, and other proponents of population genetics had to demonstrate to other scientists interested in human evolution (e.g., paleontologists, geologists, and systematists) that genetic studies of natural populations could reveal new insights into the evolution of the human species at the subspecies level—at the level of race formation. Race formation (by which biologists meant the formation of evolutionarily important groups within a species) continued to be a central focus of biological studies. Genetic studies of human populations did not replace these studies of race formation so much as they represented a new method for conducting them. Before population genetics, studies of racial differentiation in the human species had largely been limited to analyzing the fossil remains of prehistoric humans.[92] Human population geneticists, as exemplified by the Human Genome Diversity Project, promised to open new more productive avenues of inquiry by studying living human populations that they believed to be the direct descendants of the original human races. To make these connections between genetic studies of human populations, racial analysis, and the proposal of the Human Genome Diversity Project clear, below I situate population genetics in the broader history of biology.

The Emergence of Population Genetics

In the history of biology, two central questions have endured across time: Where does the diversity of life come from? Where do particular forms of life—humans, for example—come from? Famously, Charles Darwin provided a mechanistic answer to these questions: evolution through natural selection. Not God, but physical and natural processes, the originator of evolutionary theory argued, accounts for the diversity and wonder of life.

Although Darwin's theory provided a whole new way of conceiving of diversity, it left many questions in its wake: Could biological characters modified by the environment be inherited? Did mutations that resulted in sharp and discrete changes in biological characters drive evolution? Or was evolution a process of gradual change driven by natural selection acting upon naturally occurring variation? In the decades following the publication of *On the Origin of Species* in 1859, these questions generated much debate among scientists. Many accounts have highlighted the debate between Lamarckians, who believed that the genetic basis of a character could be modified directly by the environment, and Darwinians who believed that the environment could not play this kind of direct role in evolution. However, in their now classic account, *The Evolutionary Synthesis*, the systematist Ernst Mayr and the historian of biology William Provine argue that the more meaningful split existed between the "experimentalists" (most notably, geneticists) and the "systematists-naturalists" (i.e., paleontologists, geologists, and systematists) (Mayr and Provine 1980).

Experimentalists, according to Provine and Mayr, believed that evolution could best be understood by using the experimental method, a method introduced into the study of evolution by the rediscovery of Mendel's laws in 1900.[93] Mendel's famous pea experiments revealed that characters (such as smooth or wrinkled) passed on from one generation to the next in a discontinuous discrete fashion that could be codified into laws. The effort to discover these laws through experimental study became known as genetics. Provine and Mayr, as well as some prominent biologists who have reviewed this history, have noted that the early geneticists believed that the discontinuity posited by the particulate theory of inheritance (known as Mendelianism) corresponded to a discontinuity in the evolutionary process.[94] This theory encouraged what Mayr and other critics of the experimentalists have called essentialist or typological theories of evolution. Evolution, according to these theories, resulted from physical forces (most notably mutations) acting on the hereditary material (later to become known as genes). The force of the mutation produced a discrete shift in a visibly apparent characteristic or trait that led to the creation of a new biological type (Mayr 1980, 17, 13).

Mendelianists who adhered to this theory of evolution, like Thomas Hunt Morgan, perhaps the most celebrated classical Mendelian geneticist, did not believe that variation among individuals in nature could account for evolutionary change, as Darwin had argued. As Morgan wrote in 1932:

> The kind of variability on which Darwin based his theory of natural selection can no longer be used in support of that theory because . . . selection of the differences between individuals, due to the then existing genetic variants, while changing the number of individuals of a given kind, will not introduce anything new. (Morgan cited by Mayr 1980, 23)

Variation among individuals in a population, Morgan argued, could change the number of individuals of a given kind, but could not produce a change in kind—a change in species or subspecies, for example.[95] Natural populations proved too homogenous for variation *within* populations to account for species- or subspecies-level change. Only a physical force, such as a mutation, Morgan believed, could account for evolutionary change.

Mayr and Provine contrast the views of these Mendelian geneticists with those of the naturalists-systematists. Naturalist-systematists, like Mayr himself, believed that evolution could be best understood by studying variation among individuals in biological populations. Instead of a physical force (i.e., a mutation) acting upon a discrete uniform type, these scientists believed evolution resulted from natural selection acting upon variation as it occurred naturally in biological populations. In other words, the source of variation from which species arose did not come from the outside, imposed upon nature by a physical force, but rather was contained within the natural variation that existed within biological populations. To understand evolution required studying this variation as it existed in nature. Mayr and others referred to this approach to the study of evolution as the "populational approach" (Mayr 1980, 28).

On Mayr's account, as early as 1922, the geneticist Gregory Bateson called for a closer collaboration between geneticists who employed the experimental approach, and those who were systematists and naturalists. Yet it would still be a number of years before any significant steps would be taken toward a synthesis of the two approaches. Overcoming the divide, Mayr believed, would take more than importing experimental genetics into the study of natural populations, and vice versa. First, each approach would have to change. For their part, geneticists would have to incorporate more evolutionary thought into the practice and study of genetics.[96] In particular, they would have to recognize that genes do not evolve in a vacuum, but rather in a system defined by a set of interactions both between genes, and between genes and the environment. Further, they would have to recognize that this system is structured not just by events in the present, but also by a natural history as deep as a billion years. The importance of these broader systemic forces, Mayr argued, had in the past proved difficult for Mendelian geneticists to grasp, as their methodological focus had been on proximate causes: in particular, those forces that governed the translation of the genotype (defined as the inherited material) into the phenotype (defined as the expressed traits). Mayr suggested that in order for their research to become relevant to the concerns of naturalists, geneticists would have to integrate their studies of proximate causes with studies of what Mayr called ultimate causes: historical forces and events that gave rise to existing genotypes. In particular, they would have to demonstrate that ultimate causes could be explained in a manner that was consistent with the principles of genetics.

A synthesis of thought, Mayr argued, likewise would not result merely from the importation of population thinking into genetics. Classical Mendelian geneticists would first have to be convinced that a study of populations related to the study of "taxonomic units" of interest to the student of evolution (Mayr quoting Goldschmidt 1980, 17). Many geneticists, such as Morgan and Richard Goldschmidt, denied that the study of differences among individuals in a population could reveal anything of evolutionary importance. By this they meant that they did not believe that differentiation in populations related to species- or subspecies-level change. As race was the term that biologists at the time used to describe the groups that structured the evolution of inherited differences within a species (or subspecies-level change), this amounted to saying that they did not believe a study of populations related to a study of, in their terms, race. For population thinking to be taken up by evolutionary biologists, they would have to believe that it could illuminate evolution of both species, and subdivisions of species, that is, races.[97]

In short, for there to be a synthesis that would heal the divide between geneticists and naturalists, geneticists would have to prove that the complexities of evolution could be explained by studying gene action, and systematist-naturalists would have to demonstrate that the higher-order phenomena of evolution (including the formation of races) could be understood by studying populations.

A few key developments, Mayr argued, had made both of these things possible. First, William E. Castle, Alfred H. Sturtevant, and other geneticists who worked with animal breeders had demonstrated the power of selection. Artificial selection experiments, combined with the demonstrations by mathematical geneticists (such as Ronald Fisher and J.B.S. Haldane) that small selective advantages could have major evolutionary impact, helped to convince those trained in classical Mendelian genetics that selection acting on individual variation in populations could be responsible for evolutionary change (Mayr 1980, 30).[98]

Second, the discovery of interactions among genes such as pleiotropy (the effect of a single gene affecting multiple phenotypic traits) and polygeny (a character determined by several different genes) gave genes a much greater dynamism than had been attributed to them by the early Mendelian geneticists. This made it more plausible to systematists that genes could be responsible for evolutionary change.

On Mayr's account, these developments helped to break down the divides between geneticists and naturalists. Specifically, they enabled the flow of population concepts into classical Mendelian genetics, and genetic concepts into systematics. The resulting synthesis of thought accounted for the development of a new branch of evolutionary science called population genetics.

Population Genetics and Race

Population genetics is credited by many historians of biology with reforming the study of human genetics. Daniel Kevles, in his history of eugenics, gives mathematical-population geneticists such as Ronald Fisher and J.B.S. Haldane credit for overseeing the shift from the "scientific study of racial problems" to the "genetic study of human populations"—a move he claims led to the decline of "mainline eugenics" and the emergence of a respectable field of science. As Kevles argued, these two scientists "set a standard of first-class research in human genetics for scientists elsewhere to emulate" (Kevles 1985, 211).

But as we saw in the last chapter, this account of a shift from the study of "racial problems" to the study of "populations" does not capture the full complexity of the changes brought about by the advent of population genetics. Although some population geneticists did critique some concepts of race, they also argued for the importance of others. Their goal was not to replace race. As the noted anti-racist and geneticist Leslie Clarence Dunn explained:

> Race formation, whether we like it or not, is an inevitable stage in the evolutionary process by which bisexual organisms keep in harmony with their surroundings. As long as their breeding follows the normal pattern of random mating within a group of limited size, merely the limitations in the number of possible mates produces differences in gene frequency and hence leads toward race formation. (Cold Spring Harbor Symposia on Quantitative Biology 1950, 353)

Thus, Dunn concluded, "the concept of race" proved vital to the biologist's effort to create an "understanding of how our species has evolved and is evolving today; and in the light of this understanding to make some guesses about where it is tending, as a species" (ibid.). As demonstrated in the last chapter, instead of replacing race, Dunn and many other geneticists and anthropologists interested in the genetic study of human evolution merely believed that race needed to be reformed and refined in the wake of abuses by the eugenicists, white supremacists, and Nazis. Indeed, as we saw above, leading evolutionary biologists of the time believed that only by demonstrating the relevance of population analysis to an understanding of the formation of races would population genetics gain acceptance within evolutionary biology.

To demonstrate in detail these links between the study of human populations and racial analysis, and to illustrate how they bear upon the proposal for the Diversity Project, I turn to the 1950 Cold Spring Harbor Symposium, Origin and Evolution of Man. This event is particularly important in the history of race and science. Founded in 1933, the annual Cold Spring Harbor Symposium was (and perhaps still is) the most prestigious event in genetics. The 1950 gathering is of particular note as this was the year that population genetics supposedly became a mature science, bringing an end to the old racial science of eugenics (Stepan 1982, 175).[99] However, as I demonstrate below, a close reading of the transcript of this Cold Spring Harbor Symposium reveals that instead of replacing studies of race, scientists at the event intended that population genetics would refine, and even revitalize, studies of racial differences (Cold Spring Harbor Symposia on Quantitative Biology 1950, 259). Far from dropping race as an object of study, these scientists looked to population genetics for new tools that would enable a properly scientific analysis of what they termed race. In particular, participants of the prestigious symposium sought to integrate the descriptive and taxonomic tools of physical anthropology (the scientific discipline that had traditionally studied race formation in the human species) with the statistical tools of population genetics.

The 1950 Cold Spring Harbor Symposium also is of particular importance to this study as research proposed at the symposium that year strongly resonates with the research envisioned by the first proponents of the Human Genome Diversity Project. The underlying concepts, research methods, and goals of research proposed in 1950 and research proposed by Diversity Project proponents in 1991 proved remarkably similar. In the broader history of attempts in biology to understand the formation of human race through the study of "isolated" human populations, the 1950 Cold Spring Harbor Symposium is an important overture.

COLD SPRING HARBOR, LONG ISLAND, N.Y., 9–17 JUNE 1950

The 1950 symposium was attended by 129 scientists, most of whom were anthropologists and geneticists. As the geneticist Milosav Demerec observed in his opening remarks, the Symposium brought together two disciplines thought to be "most immediately concerned with man": anthropology and biology.[100] For two centuries, Demerec continued, anthropology (which studied "man" as a "social being") and biology (which studied "man" as "a living organism") had developed separately, even though both had been formatively shaped by Darwin's theory of evolution—specifically, his "finding that man is a part of nature" (Cold Spring Harbor Symposia on Quantitative Biology 1950, v). The goal of the CSH symposium that year was to work to end this division.

In this spirit of synthesis, Demerec explained, the meeting would focus on a topic of equal interest to anthropologists and biologists: "the genetic nature of traits that distinguish individual humans, or human populations such as races." In particular, three of the meeting's sessions would be devoted entirely to a consideration of the "genetic analysis of human traits," traits described as "racial traits" in the index of the volume reporting the meeting (ibid., v, xi). These sessions would be followed by an entire session devoted to a discussion of "the concept of race." As Demerec saw it:

> Many misconceptions exist in the popular mind about the nature and significance of racial variations in the human species. Considered scientifically, the problem is simple enough. Different human populations differ in the frequency of certain genes in their hereditary constitutions. The same genes that distinguish human races may also distinguish individuals within a race. Race differences are not absolute but relative. (Cold Spring Harbor Symposia on Quantitative Biology 1950, vi).

Far from viewing race as a topic that had fallen from the scientific agenda, Demerec affirmed its place in a proper discussion of human evolution and origins.[101] As many of his colleagues would agree, the problem created by the concept was not its use in science, but the misuses of it that resulted from the belief that racial differences were absolute and not relative. The meeting, as Demerec and other participants would explain at Cold Spring Harbor that year, sought to correct for these misuses by forging a new scientific study of race that integrated the expertise of anthropologists and geneticists.[102]

To this end, the meeting began with a review of how genetics and the study of evolution had developed in the last fifty years. Herluf Strandskov, a geneticist from the University of Chicago, noted that, following the rediscovery of Mendel's laws in 1900, genetic research "naturally" focused on the "individual," as the individual was the "mating unit" (ibid., 1). However, it was soon recognized that individual variation had to be understood in the context of the group in which the individual mated, and thus evolved. Strandskov referred to these groups as "natural populations."[103]

Adriano Buzzati-Traverso, Luca Cavalli-Sforza's teacher, followed Strandskov's overview with a detailed consideration of why this kind of group-level analysis of natural populations was important:

> Below visible variations, systems of genes and chromosomes are at work, which contain the variability and reveal it according to the principles of Mendelian heredity. This is why we are justified to consider *the natural population* as a unit, since individual variations must be referred to the genetic balance of the whole aggregate of individuals. The genetic structure of natural populations can not be solved only in terms of the individual variations which can be observed in the group, but must be integrated into a unitary research on changes in gene frequencies as related to the underlying breeding system [italics added]. (Cold Spring Harbor Symposia on Quantitative Biology 1950, 17)

Genetic studies, Buzzati-Traverso here argued, should not just focus on the individual, but on the group structures (i.e., populations) in which the individual evolves.

This argument resonates strongly with the argument Allan Wilson made forty years later in 1990 as he advocated for the inclusion of population studies in molecular biology projects like the Human Genome Project:

> The importance of population biology to molecular biology is also evident today as we try to understand both the forces underlying the diversity of histocompatability genes and the properties of immune-deficiency viruses. Michael Yarmolinsky makes this point when he states, "The meaning of a text, whatever its nature, derive [*sic*] from its relations to what lies outside the text. . . . Meaning is introduced with a knowledge of context, and then as an exponential function of that knowledge." (Wilson 1990, 586).

In other words, Wilson was a true intellectual descendant of Buzzati-Traverso in arguing that humans do not evolve in a vacuum as individuals, but rather in a context as members of a population.

At the time, Buzzati-Traverso viewed the emergence of "population" as a concept as "the main conceptual novelty" of the previous two decades of genetics studies on taxonomic, biogeographic, and evolutionary problems (Cold Spring Harbor Symposia on Quantitative Biology 1950,

13).[104] Recognizing that the term itself had many meanings and uses, he clarified that he was referring to a Mendelian notion of population. As Walter Bodmer and Luca Cavalli-Sforza would explain two decades later in their classic *The Genetics of Human Populations*, a Mendelian population is a "population of interbreeding individuals who share a common pool of genes, which are, of course, transmitted from one generation to the next according to Mendel's laws" (Bodmer and Cavalli-Sforza 1971, 39). For this definition they cited Theodosius Dobzhansky's *Genetics and the Origin of Species*.[105] Buzzati-Traverso also drew upon this definition, arguing that populations were "an array of interbreeding individuals" whose gene frequencies shifted over time as the result of "evolutionary factors" (Cold Spring Harbor Symposia on Quantitative Biology 1950, 13).

In providing a geneticist's definition of a population, this central innovator of the field of population genetics distinguished the notion of population used by geneticists from what he labeled the demographer's concept of population. For the demographer, Buzzati-Traverso explained, the boundaries of populations were well defined; they corresponded to some geographical, political, economic, or administrative demarcations. This demographer's notion of a population, he argued, had been employed at a time in the past when biologists were only interested in providing "anatomic, physiologic and, at most, ecologic descriptions of the various living creatures dwelling on the earth's surface." Now, though, with the advent of population genetics, the biologist had "given up the study of preserved specimens and histological microtome slides to devote himself to the analysis of how variability originates and can be transmitted from one generation to the next, by observing living beings continuously in time" (ibid.) In other words, Buzzati-Traverso suggested, the demographer's static notion of population had worked for the older *descriptive* biology that studied well-defined, preserved human remains, but it had been rendered obsolete by the advent of the new *explanatory* biology and its dynamic studies of living human populations. For this new research, geneticists had crafted a new dynamic concept of population.

Using this new concept and approach to the study of populations, Buzzati-Traverso argued, geneticists could overcome many of the problems that had earlier troubled the study of human evolution. For example, this new approach would solve one of the central problems in understanding the formation of human races: studying the distant past through the conventional method of examining paleontological and archaeological remains. As noted by George Gaylord Simpson, the geologist and curator of fossils at the American Museum of Natural History, data obtained from these remains were "highly incomplete at present," and were unlikely ever to be complete since it is "unlikely that samples of all extinct races of man and other primates now exist as undiscovered fossils" (Cold Spring Harbor Symposia on Quantitative Biology 1950, 55). To supplement the fossil record, geneticists such as Buzzati-Traverso proposed a study of the genes of present-day populations. Imagining the successful completion of a project that in some ways resembled the future Human Genome Diversity Project, Buzzati-Traverso told those assembled at Cold Spring Harbor:

> If it were already possible to draw a map of the frequencies of many genes in man over the earth's surface we would surely see some great pattern without relations to the great subdivisions of mankind, like the one of the A B O system, related to the remote ancestry of the human stock. Then we would see over this another pattern showing gene frequency densities corresponding more or less to the great races of man, this being a sign of a differentiation of long standing; and finally we would observe a very minute variegation superimposed on the two great patterns, due to the great genetic variability among interbreeding units. While the two great patterns have been determined by unknown events of the past, the variegation is due to evolutionary factors at work now; on this we must concentrate our efforts if we want to reach a better understanding of what is happening now and of what has happened in the past. (Cold Spring Harbor Symposia on Quantitative Biology 1950, 19–20)

This statement nicely draws together the understandings that linked the study of races to the study of populations. One studied current populations for what they could tell you about human evolution, including the formation of the "great races of man."[106]

Joseph Birdsell, a physical anthropologist at the University of California Los Angeles, in a session entitled "Race Concept and Human Races," further articulated the reasons for studying race formation by studying living human populations. Although highly sophisticated in its descriptive techniques, Birdsell suggested, the study of human races had suffered from its focus on the "individual." He argued that "problems of human evolution and racial differentiation are essentially population problems, and their solutions will be advanced by borrowing techniques of analysis from the vigorous field of population genetics." In his view, and the view of many of the speakers at the symposium, in order to understand basic problems of human evolution, such as the origin and formation of races, it was "an explicit imperative" that one use a "genetical definition of race"—one that posited "a population, or an isolate" as the proper unit of study (Cold Spring Harbor Symposia on Quantitative Biology 1950, 259).

It is important to restate here the relationship between the concepts of race and population used by Birdsell, as well as most CSH participants.

These scientists did not intend population to replace race (as many historians of biology have assumed). Rather they intended population to act as a category that could be used to advance the study of those biological groups in which human differences evolved, that is, races. As Laurence Snyder, a physical anthropologist at the University of Oklahoma, explained: "Speakers in the symposium have set the stage for a detailed analysis of racial traits, by outlining the broad picture of man's origin and early development, and of population units in which he now finds himself" (ibid., 159). By studying present populations, many symposium participants argued, the assembled scientists could shed light on the origin of human races. These races, according to Buzzati-Traverso, were the second "great pattern" of gene frequencies in "man." As Montagu explicitly stated, scientists wanted to study how these patterns "came into being" because they believed that races provided the primary structures in which contemporary human differences evolved (Cold Spring Harbor Symposia on Quantitative Biology 1950, 333).[107] The question was not whether, but how, to study these races. Indeed, Birdsell went so far as to say: "This symposium stands as testimony to the need for revitalizing racial anthropology" (ibid., 259).

Not only did the Cold Spring Harbor participants agree that racial differences in the human species should be studied, they believed they should be studied in as timely a manner as possible. Foreshadowing the claims of proponents of the Diversity Project forty years later, Ashley Montagu—the rapporteur for the UNESCO Statements on Race, a student of Boas, and widely credited with ending the history of race and science—told the symposium participants: "Now that the world has, indeed, become the 'same territory,' the amalgamation of all varieties of man into a single variable population is but a matter of time."[108] Thus, he concluded, "If, then, we are to understand the manner in which these geographic races came into being before this process is completed we cannot too long delay in the initiation of the necessary investigations" (ibid., 333).

The modern "amalgamation of all varieties of man" aggravated a problem addressed early on in the symposium: the lack of clear population boundaries. As Buzzati-Traverso explained, with the individual organism there is a clear "biological" boundary, but with a group, or population, "in most cases it is rather difficult to determine the boundaries" (ibid., 13–14). This problem of defining a biological population, intrinsic to any study of a population in nature, had been made worse in the human species by the advent of culture and technology. Both, Buzzati-Traverso argued, had accelerated rates of movement and mixing in the human species. Unlike the "wild goat of the Alps" that lived in small populations isolated by mountains, modern human populations increasingly lived on the same territory, not separated by geographic barriers (ibid.,

14). Thus the boundaries of the populations in which they evolved would soon be erased.

To address the challenge the mingling of populations created for the study of race formation, many at the symposium discussed the merit of studying what they called isolated human groups.[109] W. S. Laughlin called attention to studies of "geographically separated Eskimo-speaking groups." Frederick Thieme, the University of Michigan physical anthropologist, discussed the promise of studying the island "Puerto Rican population." Snyder spoke of the value of studying isolated rural groups, social classes, and religious groups. Additionally, Snyder observed that the University of Michigan human geneticist James Neel—who would later lead a research team into Brazil to collect blood samples from the Yanomama—had "extended the range of the analysis of isolates to the far corners of the earth" (ibid., 165, 25, 160).[110]

This last kind of isolate—isolates located on "the far corners of the earth"—generated much interest among symposium participants. The reason: they purportedly provided ideal conditions for population genetic research. These groups lived apart from Western civilization, with all of its technologies for mixing and moving around. Thus they had remained almost entirely stationary and isolated, and could easily be defined and studied. As Edward Hunt of Harvard explained:

> I think that although the delimitation of human breeding aggregates in many places is difficult, some parts of the world offer considerable advantages for the study of human population genetics. Among these are the islands of Micronesia. I feel that the delimitation of such breeding populations would often be easy, especially wherever the inhabitants of a single island are a locally inbreeding group. . . . In summation, Micronesia is an ideal laboratory for human biology. Its breeding populations have a cultural homogeneity and accessibility which lend themselves to the evolutionary approach. Small island populations have been invaluable since the time of Darwin in elucidating the evolution of lower forms. . . . I hope that some members of this symposium will be lucky enough to study the Micronesians. (Cold Spring Harbor Symposia on Quantitative Biology 1950, 23)

Where Darwin had studied geographically isolated finches on an island, many CSH symposium participants hoped to study what they believed to be "culturally homogeneous" isolated groups of humans—the geographically remote were considered most desirable.[111]

At the time, the desirability of this kind of research was widely recognized by those who sought to study the genetic diversity of human populations. Indeed, among those CSH participants who advocated research on isolated indigenous populations was none other than Theodosius Dobzhansky, the founding father of population genetics, and the renowned anti-racist. In an article published the very year of the CSH Symposium, Dobzhansky argued:

> It is to be hoped that data of this sort will be collected in the near future. Because of the rapid development of communications even in the most remote corners of the world, and the consequent mixing of previously isolated tribes, such investigation cannot be long postponed. In fact, it is possible that our generation is the last one which can still secure data of momentous significance for the solution of the problem of the origin of human races. (Dobzhansky 1950, 134–35)

Dobzhansky's plea followed remarks that it would be interesting to know if "American Indian populations" showed high variability in traits other than blood group traits. He hoped that these data could help shed light on the theory that human populations spread out across the Americas after the Bering Strait crossing. In sum, Dobzhansky's remarks explicitly acknowledge the link between studies of so-caled isolated human populations and studies of race formation.

Both in their proposal to study human "isolates" and in their call to conduct these studies quickly before the chance disappeared, CSH participants foreshadowed the Diversity Project. Indeed, human populations deemed isolated were the very populations that the first proposers of the Diversity Project recommended sampling. As these more recent scholars of human diversity and evolution explained in the *Genomics* article that first announced the Diversity Project: "The populations that can tell us the most about our evolutionary past are those that have been isolated for some time, are likely to be linguistically and culturally distinct, and are often surrounded by geographic barriers (Cavalli-Sforza et al., 1991, 490). As these documents indicate, both the participants at the 1950 CSH symposium and the proposers of the 1991 sampling initiative argued that the best subjects of research would be members of "isolated" human populations, and that these populations should be sampled before they "vanished."

In their early internal documents, Diversity Project organizers referred to these populations as "Isolates of Historical Interest" (Human Genome Diversity Project [HGDP] 1992, 9). Later this descriptor would be branded a moral outrage by critics who argued that researchers viewed populations as historical curiosities, and little more. Diversity Project organizers would look back on their use of the term as a naïve error (Interview C). However, in the context of the 1950 Cold Spring Harbor Symposium, one can better understand why the term was used. "Isolates" was a term frequently used by CSH participants.[112] In this context, and the scientific ways of life it represented, studying isolates for what they could reveal about human history did not represent the parochial interests of "anthropological geneticists" (a term later used to describe HGDP organizers), or bad racist science that should not have slipped by the world-renowned evolutionary biologists that led the Diversity Project. Rather, studying these populations represented a much-desired kind of research that followed in the revered footsteps of Darwin himself. Recalling Hunt's words, human isolates provided "an ideal laboratory for the study of human biology." In particular, human population geneticists and physical anthropologists at the CSH Symposium believed (as their colleagues do today) that, together, they could provide a window on the human past and the formation of human races.

RACE AND THE GENETICS OF HUMAN POPULATIONS

The connection between human population genetics and the study of the formation of human races would again be made explicit twenty years later in Cavalli-Sforza and Bodmer's *The Genetics of Human Populations* (Bodmer and Cavalli-Sforza 1971 [1999]). These distinguished human population geneticists argued that developments in the molecular and biochemical sciences had enabled scientists to study biochemical variations believed to reflect evolutionary processes—in particular, racial formation—more directly.[113] As Cavalli-Sforza and Bodmer explained:

> The biological traits that can be studied in man can roughly be classified as follows: anthropometric (measurements, external or internal); physiological, behavioral, immunological, and biochemical (especially protein differences). Only among the latter two are many characters encountered that show discontinuous variation—that is, characters whose different expressions can be attributed to the presence of different alleles. . . . It is not surprising that this is so, because biochemical variations depending on protein structure or immunological differences are much nearer to the origin of the long chain of cause and effect that starts with the gene and ends with a measurable trait. There is, thus, much less opportunity for other gene differences and environmental effects to obscure the picture. (Bodmer and Cavalli-Sforza 1971 [1999], 703–704)

In short, molecular studies of proteins were better than the "old" physical anthropologists' anthropometic study of brain cases because proteins were closer to the site of genetic change and better insulated from the whims of the environment. Cavalli-Sforza and Bodmer, however, did not believe that anthropometric traits, in particular brain case size, revealed nothing about human evolution. To the contrary, they considered the rapid increase in brain size as one of the critical developments that enabled the evolution of language and culture, and thus the evolution of the human species (Bodmer and Cavalli-Sforza 1971 [1999], 692); Cavalli-Sforza 2000).[114] However, they did believe that brain cases presented one insuperable drawback from the standpoint of a human population geneticist: one could not discern racial differentiation at the level of fossil remains.

Cavalli-Sforza and Bodmer explained their reasoning in a section entitled, "Racial Differentiation in Man." Biological races, they wrote, might be "distinctive enough in the flesh," exhibiting observable differences in "seemingly rather superficial" external traits such as a "skin color, hair texture, and nose form," however, these external racial differences could not be detected at the skeletal level. Quoting from W. E. LeGros Clark's, *The Fossil Evidence for Human Evolution*, they noted: "It may, indeed, be possible to identify a skull of a modern Negro, an Australian aboriginal, or a European, in individual cases where the racial characters are exceptionally well marked; but the variation within each group is so great that skulls of each type may be found which are impossible of racial diagnosis" (Clark quoted in Bodmer and Cavalli-Sforza 1971 [1999], 699). Thus, they speculated, it would be difficult to answer important questions about the origin of human races using only the fossil record:

> It would seem therefore that it is at present impossible to accept or reject the views that Negroes (and especially Pygmies) originate from Rhodesian man . . . that the caves near Grimaldi on the northern shores of the Mediterranean were inhabited by a Negroid race, or that Peking Man (*Homo erectus*) is the progenitor of Orientals, or many other suggestions of this kind. Moreover, the problem of the antiquity of races cannot be solved today, on the basis of paleontological evidence alone. (Bodmer and Cavalli-Sforza 1971 [1999], 699)

These important theories about the origin of human races, Bodmer and Cavalli-Sforza argued, could not be tested using the fossil record, as this record contained within it very little reliable biological information about race. Instead of the anthropometric measure of fossils that physical anthropologists had traditionally relied upon, they advocated for the study of genetic polymorphisms such as protein variants. These molecular traits, they explained, acted in accordance with Mendelian laws, showed variation between populations, were less likely to be affected by environmental forces, and thus were more likely to reflect the natural processes of human evolution. For these reasons, Cavalli-Sforza and Bodmer argued for a "genetic definition of a race" (ibid., 701).

The Genetics of Human Populations would be reprinted in 1999. In their foreword, Cavalli-Sforza and Bodmer explained that the edition had required only minor changes, as "the basic ideas of population genetics had not changed substantially over recent decades" (iii). Among the things left unchanged were the discussion of "Racial Differentiation in Man," and the call for a "genetic definition of race" (698–704).

The Human Genome Diversity Project and Race

Reading the Diversity Project in the context of the Cold Spring Harbor Symposium of 1950, and the 1971 and 1999 editions of *The Genetics of Human Populations*, clearly shows the Project's relationship to the historically well-established efforts in biology and anthropology to understand the formation of human races. As early as 1950, human population geneticists and physical anthropologists began to argue that the study of gene frequencies in so-called isolated human populations could help to illuminate the process of human evolution, in particular the formation of human races. This connection between genetic studies of isolated human populations and studies of racial formation clarifies the use in early Diversity Project texts of concepts and terms traditionally viewed as connected to Western conceptions of race. For example, according to Project documents, the provisional proposal for the Project submitted to the then administrator of the Genetics Program at the National Institutes of Health stated as a first criteria for choosing populations picking regions that were enough "isolated" that chances of "miscegenation" would be reduced. In this proposal, miscegenation, a term that figured in racially restrictive laws in the United States, served as a guiding concept for defining the proposed Project's research subjects.[115]

Further, Project organizers used terms traditionally used only to designate race to explain their central reason for doing the Diversity Project.[116] In the article that accompanied the *Genomics* editorial discussed earlier, Anne Bowcock and Luca Cavalli-Sforza explained that although the dominant use of "Caucasoid" samples in human genomics to date was understandable, these samples limited and biased the study of human genetic variation:

> Most studies of the diversity of RFLPs [Restriction Fragment Length Polymorphisms] have been made on Caucasoid samples for obvious reasons of expediency. These have been limited to the determination of gene frequencies, heterozygosities, and PICs (Botstein et al., 1980). The use of an appropriate sample of individuals from a worldwide net would increase the variation available for study. (Bowcock and Cavalli-Sforza 1991, 491)

The article went on to explain that although the authors and their colleagues had collected samples from some non-Caucasoid populations, their data were still " 'Eurocentric' " as the DNA polymorphisms they used to test these samples first had been isolated in "European populations" (ibid.).

This limitation would be reiterated in the first draft of a proposal prepared for the NIH. As Project documents record, in this proposal the existence of "Caucasoids"—a category that had been conventionally used by biologists to designate race—is not in question. It is rather the exclusive study of Caucasoids that raises concern. Instead of Caucasoids, organizers proposed studying Negritos. Negritos lived on the Andaman islands and could, they argued, represent the last ancestors of those humans who migrated from Africa to Australia. Although some labels might have been dropped or changed (here the more specific term "Negrito" is used and not the more racially inflected "Negroid"), these research interests and claims bear many striking conceptual similarities to those of the 1950 Cold Spring Harbor Symposium participants who sought to revive racial anthropology.[117]

Conclusion

As this chapter has demonstrated, the proposal for the Diversity Project emerged within the context of a long-standing—but not unchanging—effort in biology to study the origin and evolution of the human species. Since Darwin, scientists had employed an analytic concept of race to organize this research. This had not changed in its essentials by the time population geneticists proposed the Diversity Project in 1991. As the review of the report of the 1950 Cold Spring Harbor Symposium, the 1971 and 1999 editions of *The Genetics of Human Populations*, and texts of early proposals for the Diversity Project reveal, Cavalli-Sforza and other human population geneticists employed what they called a genetic definition of race. They proposed studying gene frequencies in isolated human populations—in particular, indigenous populations—not as a way to supersede studies of race, but rather as a way to study the formation of races in a manner that avoided the practical difficulties created by the modern mixing of subgroups of the human species. Proposals for these kinds of studies dated back to at least 1950 and were supported by scientists with diverse views about the biology of race—from Edward Hunt to Ashley Montagu.

All this is not to say that the conception of race used in nineteenth-century anthropology, early twentieth-century racial science, post–World War II genetics and anthropology, and 1990s population genetics were all the same. Indeed, as I have suggested in this chapter, Luca Cavalli-Sforza critiqued notions of race based on visible physical traits—a notion predominantly used by UNESCO Statements drafted in the early 1950s—and argued that this notion of race should be replaced by a concept of race defined by genes studied at the molecular level. Further, to argue that scientists continued to use race as an analytic category is not to claim that the concepts they used were unitary, unchanging, and uncontested. Indeed, as we will see in the next chapter, the proposal for the Diversity Project renewed debate between physical anthropologists and geneticists about the nature of the concept of race and its place in biological studies of human differences. The debates of the 1950s and the 1960s (discussed in chapter 2) had not been resolved, although they appeared to have been largely forgotten.

Indeed, in many ways, the Diversity Project represented a return to the research proposed during the 1950s and 1960s at forums such as the Cold Spring Harbor Symposium—only this time with a new twist. Genetic research into the origins and differentiation of the human species had in the intervening decades been transformed by the tremendous growth in molecular techniques, and the cultural and economic power of molecular genetics.[118] With these new and powerful techniques, the scope of research imagined by population geneticists had grown from individual studies to a worldwide survey, while the importance of intellectual collaboration with anthropologists had receded. Equally consequential, the emergence of the indigenous rights and civil rights movements had transformed the practices and politics of defining and studying populations. The result was that old debates among geneticists and physical anthropologists about the concept of race and the study of human populations resurfaced—this time with intensified energy, new actors, and increased stakes.

Chapter 4
Diversity Meets Anthropology

Although at the inception of the Diversity Project organizers did not consider questions about the meaning of race and its role in structuring their initiative, this would all begin to change as the first criticisms of the Project emerged just months after the call for the survey appeared in *Genomics*. At this time, *Science* magazine published a letter from Mark Weiss, then director of the Physical Anthropology Program at the National Science Foundation. In this letter Weiss called into question the lack of involvement of anthropologists in the planning of the Project, arguing that these scientists possessed invaluable expertise that would be required at every stage for the initiative to succeed.[119] Project organizers responded promptly to Weiss's concerns. Within months they added a physical anthropologist to the initiative's planning committee and put this scientist in charge of a planning workshop that sought to enroll anthropological expertise. Yet far from calming worries, this effort only raised new concerns. Indeed, by the spring of 1993, not only did some physical anthropologists worry that the Diversity Project would exclude them, they charged that the Project threatened to revitalize scientific racism by imposing colonial categories of race onto the study of human diversity. By drawing upon critical studies of expertise in science and technology studies, this chapter seeks to explain why. Specifically, the chapter reinforces the insight that expertise does not exist objectively as something to be enlisted; rather it is constructed through a complex set of often-contested negotiations that interweave conceptual and political issues (Jasanoff

1990, Jasanoff 1992, Hilgartner 2000). At stake in these negotiations are not just questions about how nature should be understood, but also about how societies and cultures should be ordered, for whom, and to what ends (Epstein 1996, Wynne 1996).

Perhaps nowhere is the entanglement of expertise with social order and power more evident than in the human sciences. As the French analyst of power and knowledge Michel Foucault demonstrated through his studies of madness, the clinic, and the prison, these sciences do not explain the human so much as they constitute the human as an object that can be known, and thus a subject that can be governed in novel ways (Foucault 1973, 1975, 1976). With their advent, no longer does power rely on the negative techniques of taking away the rights, and even life, of the subject. Now it can act in far more precise and invasive ways through the productive techniques of creating a subject who seeks to maximize his or her potential for life through knowing him- or herself. In this new age, more than the rule of law, rules that govern what can and cannot count as knowledge determine what human lives can and cannot be lived, a power Foucault named biopower (Foucault 1976).[120] Given the centrality of knowledge about "man" to modern regimes of power (a relation Foucault highlighted through use of the contraction "*power-knowledge*"), not surprisingly the ideas and practices of the human sciences have often proved controversial.

Diversity Project organizers' efforts to include anthropologists faltered and generated concerns about racism as they failed to recognize this entanglement of the human sciences with modern formations of power. Organizers of the initiative merely agreed with Weiss's assessment that to succeed, the sampling initiative would need to include other scientists—namely, anthropologists. What they (and Weiss) did not adequately recognize was that anthropology did not exist as a sufficiently unified body of expertise that could be unproblematically enlisted. Rather, it proved a fractured discipline in the midst of a much broader and historically embedded debate that raised fundamental questions about the proper methods and objectives of its practitioners. Should the anthropologist study biology or culture or both? If both, how should the study of biology and culture be integrated? What role could genetic research play in this integration?

These questions about the proper goals and methods of anthropology connected to fundamental questions about the nature of power-knowledge. What does it mean to possess knowledge of the human (*anthropos logou*)? Who can speak for humanness? What is to count as a fundamental human science? Human population genetics? Physical anthropology?

Cultural anthropology? If human population genetics emerged as a dominant human science, would human experience be reduced to genetic processes? Would this result in novel forms of racism and colonialism? Who even has the power to define racism and colonialism in an age where the very definition of race and human differences had become subject to the expert techniques of genomics?

By drawing on a framework that presumed the prior existence of expertise, and opposed this expertise to reprehensible social forces, organizers and critics alike prevented an interrogation of these more fundamental questions about the mutual constitution of expertise and power. Organizers and critics alike simply argued that studies of human genetic diversity that employed proper scientific methods should move forward, and those that propagated racism and colonialism should desist. All assumed they knew the difference between the former and the latter. Consequently, deeply entrenched and contested questions about the nature of knowledge about the human, and the relationship of this knowledge to modern techniques of power, arose in explosive moments, but never received systematic or sustained attention. Thus, organizers largely averted the work needed to craft practices and ideas that could stably order human diversity in an age of intensifying genetic research.

Questioning the Boundaries of the Human: Early Tremors of Debate

Questions about how to define and order human beings for the purpose of sampling their genetic diversity did arise early in the efforts to organize the Diversity Project. In retrospect, however, the reported differences between Luca Cavalli-Sforza and the UC Berkeley biochemist Allan Wilson over sampling strategies in the spring of 1991 proved to be one of the most uncontroversial episodes in the history of the Project. In almost every way these two original scientific leaders of the Project agreed: a worldwide survey of human genetic diversity should be undertaken; this survey would involve collecting and preserving DNA samples from "vanishing indigenous peoples" (Roberts 1991a, 1615). Their only disagreement revolved around how to sample indigenous DNA. Cavalli-Sforza advocated a population-based approach. This method would involve collecting blood and other human tissue samples from 50–200 "genetic populations" chosen using a variety of criteria (Roberts 1992, 1204). First, language. As the report from the first planning workshop for the Diversity Project would later explain: "Groups with distinctive languages can often be regarded as a genetic population." In addition, the report indicated that populations that met the following criteria would be of greatest interest to the Project: "mostly rural"; no "genetic admixture"; "populations that are vanishing because of mortality, migration, admixture, etc., and that are potentially important for historical genetics"; "representative of the world, specifically, of the world before the expansion of present dominant groups" (Human Genome Diversity Project 1992a, 5). Cavalli-Sforza believed that blood and other human tissue samples collected from these populations should be transformed into cell lines. Nuclear DNA distilled from these cell lines, he argued, would prove enormously valuable for studies of the evolutionary history and population dynamics of the human species.

Wilson challenged the validity of this population-based approach. Some genetic variants, he argued, occurred only within a small geographic radius of fifty or one hundred miles. Thus, if the Project only sampled populations defined by cultural and linguistic criteria, as Cavalli-Sforza proposed, much genetic variation would be missed. To solve this problem, Wilson proposed sampling individuals on a geographic grid. Instead of collecting samples from populations, this approach entailed "collecting DNA from aboriginal peoples at more or less evenly spaced locations around the world."[121] Collecting in this manner, he explained, would enable researchers "to be explorers, finding out what is there, rather than presuming we know what a populations is" (Roberts 1991a, 1615).

Wilson's dream of discovering natural truths by employing methods untainted by social and cultural categories, however, also encountered problems. To begin with, critics of the grid approach worried that representatives of "aboriginal populations" (however one ended up defining these populations) would not exist at fixed intervals across the globe (for example, every fifty or one hundred miles). As a noted population geneticist at an East Coast U.S. university observed:

> I mean just going out and sampling now in a grid fashion, for example, regardless of what you know, you may just—if you go through the middle of Zimbabwe, or some other sort of thing, and you just sample randomly—you may pick up a tourist from Ohio. I mean, you know, if you just drop a dart down. (Interview Y)

Although social and cultural categories might not perfectly map onto meaningful genetic ones, in his opinion they still could serve as useful tools for ordering the sampling initiative.[122] And, indeed, Cavalli-Sforza agreed with this view. He did not disagree that the Project should be sensitive to geographic differences within groups. Rather, internal memos document that he only argued that ignoring the "ethnic background" of groups would lead to missing discontinuities that resulted from "cultural" variation, historical accidents, and endogamous tendencies within groups defined by ethnicity.

In the end, those in attendance at the first planning workshop for the Diversity Project decided to sample twenty-five individuals from four hundred populations for cell line transformation. In addition, they decided the Project should collect a "large" number of "non-immortalized" samples from individuals (Human Genome Diversity Project 1992a, 4). Collecting these additional DNA-only samples would enable the Project to be more inclusive and to accommodate Wilson's viewpoint. The additional samples would allow the Project to test its assumptions about the definition and reality of populations.

However, despite this settlement, questions about the Project's sampling strategy would continue to trouble Project organizers.[123] Although perhaps of little consequence when framed as part of a disagreement between Cavalli-Sforza and Wilson, when placed in their historical context these questions gain a larger significance. In this broader context, connections between the Cavalli-Sforza and Wilson debate and deliberations at the 1950 Cold Spring Harbor Symposium become clear. As described in the previous chapter, many geneticists and physical anthropologists at the CSH symposium argued that the appropriate unit of analysis for genetic studies of human diversity would be the population.[124] However, questions arose over how to define this unit of analysis. Scientists such as Buzzati-Traverso, one of Cavalli-Sforza's teachers, argued that this definitional problem could be solved by studying human isolates—well-bounded human groups isolated by geography, and unified by a common language and culture. Forty years later, Cavalli-Sforza made a similar argument, and like his population geneticist predecessors, he encountered resistance. Echoing earlier critiques by clinalists (such as the physical anthropologist Frank Livingstone), Wilson argued that human diversity might not map onto recognized population boundaries, and thus the use of population categories should be suspended.[125]

In other words, the differences between Cavalli-Sforza and Wilson connected to long-standing debates in anthropology and genetics about the origins of human diversity: Where does diversity come from? Individuals or groups? If diversity is a group-level phenomenon, then how should groups be defined, by whom, and for what purposes? As we will soon see, these questions proved so persistent and intractable because they related to entrenched and contentious debates about how the human could be known and governed.

WHO CAN KNOW THE HUMAN?: THE DEBATE EXPANDS

In addition to questions about how to define the object of research, questions about who could be the knowing subject arose following the publication of the first *Science* article about the Diversity Project in June 1991. For this article, Leslie Roberts, a reporter for *Science*, sought out "an anthropologist's perspective" on the Project.[126] Viewed within the broader historical context laid out in the previous two chapters, Robert's impulse to seek out such a perspective made sense: anthropology had traditionally been the discipline that sought to study the origins and diversity of the human species; thus a science journalist might reasonably consider an anthropologist an important expert to evaluate the Project. However, Robert's explicit effort to seek out an anthropological perspective did represent a moderate change in the development of the Project. Up until the publication of her article, Project organizers had focused most of their attention upon gaining the support of geneticists. For example, the article announcing the Diversity Project appeared in *Genomics*, a journal read mostly by geneticists, not anthropologists. Furthermore, no recognized anthropologist sat on the Human Genome Organization committee that drafted the first proposal for the Project.[127] As a result, at the time the first article about the Project appeared in *Science*, most anthropologists had not heard about it.

Diversity Project organizers' efforts at the very earliest stages of planning to seek out the support of geneticists, not anthropologists, reflected an approach that defined creating a sound and fundable scientific agenda as the first step in developing a science project. According to this approach, only after the science had been determined did it make sense to focus on mobilizing the resources needed to conduct the research.[128] As project documents and internal memos indicate, at this early stage in the planning process, organizers viewed anthropologists as among those resources (those who could help generate pedigree information and collect blood samples), and thus viewed questions about the role they would play as secondary. However, as would soon become clear, this would not be a viable strategy. Since physical anthropologists were recognized as experts on the diversity and evolution of the human species, their absence from the early planning stages would spark concerns about the Project's scientific legitimacy.[129]

This became evident following the publication of Roberts's article. The article sparked a flurry of responses from the physical anthropology community. These began with a letter to the editor of *Science* written by the then director of the Physical Anthropology Program at the National Science Foundation, Mark Weiss.[130] In his letter, published on 27 September 1991, Weiss argued that scientists in circles broader than the genomics community had recognized "the need to document human variability before it disappears." Indeed, he added, "surveying of human genetic diversity is also a central goal of the Physical Anthropology Program at the National Science Foundation." Echoing claims made in the *Genomics* article that announced the Diversity Project, Weiss explained the rationale for this goal: "As isolates become admixed or otherwise disappear, the genetic information they can provide about population history and evolutionary processes also vanishes" (Weiss 1991, 1467).

Given these common interests, Weiss argued that Project organizers should involve physical anthropologists in the proposed sampling initiative. He challenged Roberts's assessment that the relatively small size of the NSF Physical Anthropology Program would prevent collaboration on what Weiss referred to as a " 'mini-genome' " project. To the contrary, he argued that the Program already supported such collection efforts, and although it could by no means support the full scope of the sampling project proposed by Cavalli-Sforza and his colleagues, it could assist in a "coordinated effort" through "competitive awards" to "the community of scientists interested in human genetic diversity" (ibid.).

Not only could the field of physical anthropology offer monetary support, Weiss argued that it could offer necessary expertise. Indeed, he believed that a study of human populations could not be done without the expertise of physical anthropologists. These anthropologists had experience working with populations and constructing rigorous data sets from which one could derive sound conclusions. Without the benefit of their experience and expertise, Weiss warned, genetic studies of human populations might not take into account the complex set of cultural factors that shape human diversity. Without the benefit of the rich understanding of human groups that physical anthropologists possessed, the Diversity Project's object of study, the human, might be reduced to a collection of gene pools defined by mechanistic laws (Interview with author 1999).[131]

Finally, Weiss believed that physical anthropologists would ensure that the sampling initiative would result in a resource valuable to all who studied human genetic diversity. Although the article announcing the Project did present the sampling initiative as an effort that would interest a wide range of scientists, its publication in the relatively narrow circulation journal *Genomics* privileged some scientists' access to the Project over others. Further, the *Genomics* article focused on the particular kinds of questions that would be answered (questions about human evolution and origin) rather than on designing a project that would represent the genetic diversity of the human species. The fact that the initiative had been formed around research questions of interest to a particular group of scientists would mean that it would likely exclude the research interests of others.

Instead of focusing on particular research questions, Weiss contended that the initiative should seek to solve what he called "the empty matrix problem" (Interview with author 1999). He borrowed this term from Cavalli-Sforza and used it to refer to the problem created by the haphazard collecting that had marked the previous fifty years of sampling human populations: samples had been tested for genetic markers of interest to a particular researcher rather than a standard set of markers that would be of interest to all. The result was that a matrix created by positioning "population" along one axis and "genetic loci" along the other would contain many blank spaces. According to Weiss, solving this problem by creating a full matrix would benefit all researchers. This would require expanding the Project's conception of the object of sampling from so-called vanishing populations to populations that would lead to a representative sampling of human genetic diversity.

As Weiss's arguments make clear, what had begun as a debate about the appropriate object of study could not be disentangled from a debate about the relevant knowledge base. Questions about who would be sampled entailed questions about who had the expertise, and thus the authority, to sample. Understood in the broader context of the history of efforts to study human diversity at the genetic level, this expansion of the debate to include questions about the construction of expertise should not be surprising. As early as 1950, at the Cold Spring Harbor Symposium, students of human evolution had argued that studies of human genetic diversity would require the expertise of anthropologists as well as geneticists.

The first proponents of the Diversity Project were receptive to Weiss's suggested amendments. Indeed, very soon after the publication of Weiss's letter in *Science*, conversations began between Luca Cavalli-Sforza, Mark Weiss, and Kenneth M. Weiss, the physical anthropologist whom Roberts had contacted for her June 1991 *Science* article.[132] These conversations led to meetings on 30 November–3 December 1991, at Stanford University at which Cavalli-Sforza, Marcus Feldman, Mary-Claire King, and Kenneth Weiss wrote a proposal entitled *Human Genome Diversity: A Proposal for Two Planning Workshops and a Conference*. Mark Weiss then coordinated support for this proposal from programs at the National Science Foundation, the National Institutes of Health, and the Department of Energy. For the first time, an anthropologist had taken the lead in organizing a part of the initiative.

Anthropologists would again take the lead when Kenneth Weiss organized the second planning workshop supported by the NIH/NSF/DOE conference grant. This workshop, originally named the Anthropology Workshop, convened on 29 October 1992, and brought geneticists from mostly health fields together with anthropologists interested in ethnic diversity and evolution—about fifty in all (Interview V). The goal of the meeting was to create a sampling agenda for the Project. Organizers of the workshop explained that "a special effort [would] be made to identify populations threatened with imminent cultural or biological extinction or dissolution." This strategy echoed the one proposed in the *Genomics* article. However, the workshop would also reflect new sampling strategies introduced by anthropologists. As stated in the workshop proposal, "there are other important anthropological reasons to sample in other ways as well" (ibid., 9). For example, samples were to be collected that would lead to the creation of a "representative sample" of "entire continents" (Roberts 1992b). Deciding what exactly it meant to create a representative sample, and how this new goal could be integrated with the original goal of collecting "disappearing" populations, would later be a focus of debate. At the moment, however, it is only important to note that the addition of these other sampling strategies marked a slight but significant change in the stated goals of the Project that accommodated the anthropological vision described by Mark Weiss.[133]

In addition to introducing new objects of sampling (e.g., representative populations), the second planning workshop confirmed the importance of adding new research perspectives (i.e., experts). As the report produced from the Anthropology Workshop explains, "sound anthropological expertise" would be *necessary* and not optional for the Project: "The constraint to use this document as a framework for sampling guarantees that the Project will remain an open one, that is properly determined by *sound anthropological expertise* [italics added]" (Human Genome Diversity Project 1992b, 4). According to the report, all regional groups (e.g., the North American group, the Indo-Pacific group, etc.) agreed that the Project would need to have "actual sampling done as part of full and *legitimate anthropological* (linguistic, health, social, etc.) studies of the group, by investigators trusted by and familiar to the studied population" [italics added] (ibid., 5). Following the Anthropology Workshop, the importance of anthropological expertise would continue to gain prominence. Indeed, in a Senate hearing held in April 1993, organizers claimed that part of what made their effort unique and different from previous sampling efforts was that the work of the Project would be carried out by "scientifically-trained anthropologists" (Committee on Governmental Affairs 1993, 27).

With the addition of Kenneth Weiss to the planning process, his appointment as organizer of the Anthropology Workshop, and the calls to enroll scientifically trained anthropologists, those who first proposed the Diversity Project might have understandably believed that they had responded to Mark Weiss's call for the inclusion of physical anthropologists. However, as they would soon discover, important questions remained about the meaning of inclusion and the nature of anthropological expertise. These questions, in turn, would lead back to the debate about who should be sampled.

WHO IS THE ANTHROPOLOGIST—ASSISTANT OR EXPERT?

Shortly following the addition of Kenneth Weiss to the planning committees, Project documents began to acknowledge that work would be required to bring other anthropologists "on board."[134] Organizers realized that the human sciences remained fractured along disciplinary lines (despite calls for the integration of anthropological and genetic expertise at the CSH Symposium decades before). To overcome this divide for the purposes of conducting the Diversity Project, the geneticists who proposed the Project anticipated a need to publicize their sampling initiative to "anthropologists" by publishing articles about the Project in "world anthropological journals" as well as other outreach activities (Human Genome Diversity Project 1992b, 4). As part of these outreach efforts, in April of 1992 Cavalli-Sforza gave a plenary address at the Annual Meeting for the American Association for Physical Anthropologists (AAPA). In this address he described "a program for a genetic survey of the human species" to several hundred meeting attendees, all presumably interested in the issue of human variation (Interview T).[135]

Although intended to generate the support of physical anthropologists, Cavalli-Sforza's talk only sparked new concerns. Among these many concerns would be the worry that geneticists working on the Diversity Project would in effect place physical anthropologists in subservient roles. As one member of the audience recalled, Cavalli-Sforza's request for help in the collection of samples for the genetic survey elicited mutterings of the kind: "Oh, sure, and then turn the samples over to the 'real scientists' to study . . ." (Interview W).

Similar worries would arise as population geneticists started contacting physical anthropologists about helping in the collection process. As one researcher from an American university recalled, he was just going about the normal course of doing his research when he began to hear about the Diversity Project from colleagues. In particular, he remembered his response to a geneticist colleague who was at the time involved in the planning of the Diversity Project and who solicited his advice on how to sample:

> This is not how we do anthropology. This is not how we establish relationships with groups. You don't just take samples from groups. . . . [Y]ou don't just take samples from people and they never see you anymore, you never hear from them and you do whatever you want with them. (Interview L)

At first, this anthropologist attempted to help reform the Project by offering his advice. However, when many of the bench scientists associated with the Diversity Project began to ask him to collect samples for them, his commitment to helping waned. The request to sample, he believed, derived from geneticists' desires to not "have to be bothered with the people" (Interview L). They sought to study genes, not cultures and people, he argued. People served as mere vessels for genes. Culture served as a way to locate genes of interest. What he perceived to be the Project's instrumental approach to the study of human populations troubled him intellectually and ethically.

Sometimes instead of samples, this anthropologist recalled, geneticists asked for contacts: "They would say to me, we don't want you to go to Africa, we just want your contacts" (Interview L). He received these requests through phone calls and at meetings in 1991 and 1992, interactions he likened to those of a "procurement officer" (Interview L). No longer positioned as an expert, he felt himself demoted to assistant.

Despite an indicated commitment to collaboration as colleagues, these experiences demonstrated that in practice a hierarchy remained: geneticists would design what they viewed to be the science while physical anthropologists would help to facilitate it. This hierarchy in the subjects of knowledge production connected to a hierarchy in the objects of knowledge: the physical anthropologist's interest in the biological and cultural processes that shape human populations was subordinated to the geneticist's interest in discovering the laws of inheritance and evolution; culture was not a central object of study so much as it served as a tool that could be used to identify gene pools.

This instrumental approach to anthropology would not just be noted by physical anthropologists. It would also manifest itself in Project documents and internal memos. These documents and memos confirm that in the very early stages of planning, organizers imagined anthropologists as a group of scientists whose experience working with populations in particular regions might prove useful to the sampling efforts. They did not view them as colleagues who would help at every stage of the design of the Project.

Traces of this first understanding of anthropologists as potential assistants, and not collaborators, remained in the report of the Penn State Anthropology Workshop. This report distinguished between "Organizers" and "anthropologists" in passages such as the following: "The intention of the Organizers is that the actual data collection will be done not by them, but by investigators (anthropologists, health workers, and so on) who have regular working relationships with populations" (Human Genome Diversity Project 1992b, 4).[136] In other words, although authors of the workshop report recognized anthropologists as necessary bearers of expertise, they also continued to describe them essentially as collection agents. In short, the role of assistant imagined for the anthropologist would not disappear; a role of collaboration would just be added to it. This is made explicit in passages like this one:

> Generally, the organizers will attempt to continue to publicize the need for *assistance and collaboration* as widely as possible in world anthropological journals and the like. Recommendations will be changed as such information is received [italics added]. (Human Genome Diversity Project 1992b, 4)

Here the role of assistance joins and does not replace the role of collaboration.

This understanding of anthropologists as those who could help human population geneticists with their research was not new. Rather, it followed from the previous experiences of population geneticists who had collected human tissue samples from indigenous populations. For example, when in the 1960s Cavalli-Sforza traveled to Africa for the second time to sample, he contacted Colin Turnbull, a social-cultural anthropologist from Scotland whom he had worked with in Zaire. As Cavalli-Sforza later recalled, his first expedition to Africa suffered for lack of working relationships with anthropologists:

> That first trip was a pilot investigation. The major objective was to locate Pygmies, since their whereabouts were only vaguely known, and then to convince them to give us blood samples. . . . Local farmers were our initial resource for locating Pygmies. They were the only ones who knew the trails to the Pygmy camps in the forest. It would have been preferable to have anthropologists working for considerable periods of time in different places before we arrived, but this was hardly compatible with our organization and financial support. (Cavalli-Sforza 1986, 3)[137]

Following this expedition, Cavalli-Sforza contacted Turnbull, who was then studying the Pygmy people living in the Ituri Forest in the Congo.[138] Turnbull subsequently joined Cavalli-Sforza twice on his research trips—once in 1969 and once in 1971. Most significantly, in 1971 Turnbull hosted Cavalli-Sforza and his team for a week in Epulu, Ituri, Zaire. On this occasion Turnbull provided support for the collection of blood samples, and demographic and medical information from the local Mbuti Pygmies. It was rare that Cavalli-Sforza had the chance to work with an anthropologist. He recalled that he very much enjoyed the help and knowledge that Turnbull provided.[139] Indeed, he regretted that the time spent with this reknowned cultural anthropologist was limited.

James Neel, a University of Michigan geneticist who founded the first department of human genetics and led the first expeditions to sample indigenous populations, also recalled the importance of cultural anthropologists. In an effort very much like ones envisioned by Cavalli-Sforza and proponents of the Diversity Project during the 1960s, Neel led expeditions to Brazil to study the gene pools of the Yanomami.[140] The presence of a cultural anthropologist, Neel later observed, proved important to the practicality of this research:

> We would try to pick an Indian village maintaining intermittent contacts with one of the outposts of the Brazilian Indian Protective Service. It would have to be one of the outposts with a landing strip, so we could move quickly when we had our perishable samples. First priority would go to any such villages in which a cultural anthropologist had already worked; we would try to persuade him to join us in the field. (Neel 1994, 122)

So when one of Neel's collaborators, Francisco Salzano, wrote to Neel to tell him that he had discovered such an anthropologist, the Harvard cultural anthropologist David Maybury-Lewis, Neel immediately arranged to meet with Maybury-Lewis.[141] Ultimately the group Maybury-Lewis studied, the Xavante, became the first group that Neel and his collaborators sampled for their pilot study (Neel 1994, 123–25).

In the context of these prior experiences, moments in Diversity Project documents and internal memos that assume that anthropologists might help with the collection of samples make sense. However, this understanding of anthropologists as people who could aid with the pragmatic and practical dimensions of research was no longer viable in the early 1990s. The incorporation of genetics training into the physical anthropology curriculum during the 1970s and 1980s led many physical anthropologists to view themselves as experts in the field of human genetic diversity research. Notwithstanding Carleton Coon's fears that genetics threatened to turn anthropology into a narrow and reductionist science (as noted in chapter 2), many physical anthropologists (as well as other life scientists) had by the 1990s incorporated genetic analysis into their research. Indeed, the first chair of the Project's North American Regional Committee, Kenneth Weiss, is quoted in *Science* as stating that genetic research is "the future of the whole area." "I am not saying genes are everything," he explained, "but [these techniques] are a major tool for reconstructing history, taxonomy, and phylogeny" (Roberts 1991b, 1614).

In short, by the 1990s physical anthropologists viewed themselves not just as scientists who could assist in genetic studies of indigenous groups, but also as scientists who could themselves organize and conduct the research. Indeed, some believed that they, and not geneticists, should lead the sampling initiative. Defining the object of study and collecting samples, they argued, would require both biological and cultural expertise, expertise possessed by physical anthropologists, not geneticists (Interview

W).[142] Thus, not surprisingly, these anthropologists challenged the idea put forward by some geneticists that they would be assistants, and not primary collaborators, in the proposed sampling initiative.

A Tale of Two Meetings

As many of these early interactions with physical anthropologists indicated, even if organizers of the Diversity Project agreed in principle that anthropologists should be included as experts, achieving this goal would require overcoming historically embedded practices that placed anthropologists in the role of the geneticist's assistant. However, reworking these prior relationships would not be the only challenge human population geneticists faced as they sought to build ties with anthropologists in the 1990s. As organizers of the Diversity Project soon discovered, even if the Project raised no questions about the status of anthropologists, the effort still tangled with long-standing questions within the discipline of anthropology about the nature of anthropological expertise.

Early on, the geneticists who first proposed the Diversity Project showed little awareness of these debates. Instead, they merely referred to anthropologists as if they comprised a cohesive group whose support they could enroll. So, for example, articles about the initiative that began to appear in physical anthropology journals argued that studies of human genetic variation would have value for traditional areas of interest to anthropologists and would reveal answers to "core problems in anthropology" (Kidd et al., 1993, 2–3). At this point in planning the Project, organizers assumed that anthropology was cohesive enough as a discipline for it to have identifiable core problems and traditional areas of interest.

That this might not be the case would be made evident by the Wenner-Gren International Symposium entitled Political-Economic Perspectives in Biological Anthropology: Building a Biocultural Synthesis. In a juxtaposition its co-organizer Allan Goodman would later describe as "mind boggling," this symposium was held in Cabo San Lucas, Mexico, the very same week the Diversity Project's Anthropology Workshop took place at Penn State (Goodman 1992).[143] Like that meeting, the Mexican symposium sought to create a dialogue between those who studied human diversity from cultural and biological perspectives.[144] In particular, meeting attendees sought to overcome the divide between physical and social-cultural anthropology. As the introduction to the symposium volume explained, the object was to "[correct] years of fragmentation of the many subfields of anthropology, and to [correct] the loss of the important focus on the holistic study of the human species that has been anthropology's distinct contribution to the human sciences" (Goodman and Leatherman 1998, xix).

Both the participants in the Wenner-Gren Cabo San Lucas meeting and the Diversity Project Anthropology Workshop shared the goal of synthesis. Indeed, this goal was one that the historian of anthropology George Stocking had argued for in his classic *Race, Culture and Evolution* (1968). As explained in chapter 2, Stocking argued that race arose as a concept in the human sciences in the nineteenth century when static physical theories of "man" usurped dynamic environmental or cultural theories. The ascent of race, he contended, accompanied the rise of European colonial power. To overcome what Stocking viewed as this encroachment of ideology into science, he argued that dynamic concepts, such as that of culture, would have to be restored and reintegrated with physical theories of the human.[145]

But if many practitioners and observers of anthropology had called for a synthesis of biological and cultural expertise, then why, after decades of advocacy, had it never been achieved? Further, why were there two separate conferences that sought to integrate biological and cultural expertise, and why were most of the attendees who attended one unaware of the other?

These questions point to central tensions in the discipline of anthropology that would be important to the Diversity Project debates. In particular, they call attention to struggles over the meaning of biology and culture. Although all the anthropologists involved in the Diversity Project debates agreed that human differences were both biological and cultural, they did not agree on what constituted these two domains, or on methods for studying them. In particular, those who supported the Project and those who critiqued it often operated with two different definitions of culture. On the one hand, the original proposers of the Diversity Project operated with the notion that culture consisted of customs that were transmitted across the generations and helped to maintain biological integrity.[146] This concept of culture applied best to the "isolated" indigenous groups that Cavalli-Sforza and his colleagues had originally hoped to sample. The anthropologists at the October 1992 Diversity Project anthropology meeting at Penn State at least partially drew upon this notion of culture as they selected populations for sampling.

Anthropologists gathered at the Wenner-Gren symposium in Cabo San Lucas, on the other hand, defined culture as composed of structures that did not maintain the purported integrity of biological groups, but rather transformed these groups. They focused not on what populations maintained and passed on from generation to generation, but rather on dynamic processes of change. Further, they critiqued studies that identified "isolated" populations whose genetic integrity had been preserved by cultural integrity. Participants at the Wenner-Gren symposium argued that these notions of culture and isolated populations were based on the false belief that the "proper order of human life" would be found in intact cultures that existed "outside human society." This belief was particularly pernicious, they argued, because it had led scientists to abandon research on "major populations" (such as "African Americans") in favor of studying "seemingly isolated, traditional cultures where natural science theories had the most explanatory power." The result was that "the most significant social and biological problems of the species" dropped out as research priorities in biological anthropology (Blakey 1998, 382, 385).

By reintegrating biological and cultural anthropology, Cabo San Lucas symposium participants sought to make anthropology once again relevant to the concerns and problems of humanity. They specifically hoped that by merging ecological and evolutionary theories of adaptation with political-economic theories of power and social relations they would create new knowledge that could help reduce the burden of disease and the spread of conflict in a world defined by the ills of global capitalism: displacements of populations, environmental contamination, new and reemerging infectious diseases, and an increase in ethnic conflict (Goodman and Leatherman 1998, xix–xx, 25).

As the views of the participants at these two meetings illustrate, although bringing the study of biology together with the study of culture had been a long-standing goal in anthropology, different members of the field had different understandings of what this would mean. No agreement existed even on the most foundational issues, such as the definition of culture. Given this, the goal of synthesis raised the following questions: Whose accounts of biology and culture would be synthesized? On what authority could culture and biology be defined? Anthropologists who in language articulated a similar goal, in practice attended different meetings and adhered to different methods for studying human diversity. Whose meetings and methods would gain prominence and legitimacy?

Expertise and Power

Questions about the definition of culture and the proper methods and objects of study in anthropology were not new. They had been preceded by decades of debate and struggle within the discipline of anthropology that followed the official demise of colonialism.[147] As many anthropologists and scholars of anthropology have noted, much of the history of anthropological thought is complicit with European imperialism and domination (Kuklick 1991, Harrison and Harrison 1991, Stoler 1995).[148] For example, British patrons provided the support needed for anthropology to emerge as a "recognizable field of study" precisely because they believed that anthropological knowledge could be used to manage the people and land of the British Empire (Kuklick 1991, 6, 25). Given this legacy, not surprisingly questions arose about the status and legitimacy of anthropological ideas and concepts once colonialism, in its traditional form, crumbled following World War II. Indeed, the very role of the anthropologist fell into question.

As the social historian of anthropology Henrika Kuklick has argued, government officials in Britain hoped that anthropologists would still play a service role in this postcolonial age and help turn former colonies into independent nations. Many leading anthropologists of the time, however, rejected this role. "Scientists," they argued, "do not do commissioned research." The denigration of the role of service—a role that anthropologists had not only accepted, but also once embraced—marked a profound transformation in the discipline. Rather than remaining "practical men," anthropologists sought to constitute themselves as professional academics. As Kuklick explains, the "business of anthropologists became the training of other anthropologists to occupy professional posts" (ibid., 14). This is not to say that anthropologists retreated to a place outside of the governmental and social realm (this, she argues, would have been impossible). Rather, she suggests, connections to these realms became suspect within the field.

During the 1960s, worries about the links between anthropology and government were aggravated by alleged links between anthropologists and U.S. military officials. For example, Project Camelot, the suspected use of anthropologists by the U.S. Department of Defense to help in insurgency and counterinsurgency activities, left the discipline asking questions about its proper goals and (ab)uses (Horowitz 1967).[149] By the early 1970s, worries about access to field sites, uncertainty about the proper objects of research, and concerns about the ethics and methods of fieldwork precipitated what some called a crisis of identity that necessitated nothing short of the reinvention of anthropology (Stocking 1982 [1968], xii).

What exactly would emerge as the new goals of anthropology during this era of crisis became a point of debate. In particular, many raised questions about how to reconstitute the discipline in such a way that its historical legacy of supporting colonialist and imperialist projects would be broken. Early in the century, Franz Boas had argued that an anti-imperial and anti-racist anthropology would require approaches that emphasized the cultural dimensions of human differences.[150] Inspired by his school of research, a new field of cultural anthropology emerged. Over the coming decades, cultural anthropology grew ever more institutionally and intellectually separate from what would become known as physical anthropology.

The points of difference between cultural and physical anthropology, however, would for many years remain elusive to many practicing anthropologists. Indeed, those now labeled physical anthropologists argued that they too had been inspired by Franz Boas and also placed an emphasis on the study of culture (O'Toole 1998). Like the anthropologists and geneticists at the 1950 Cold Spring Harbor symposium, these physical anthropologists argued that a study of biological processes should be integrated with a study of culture. Over the ensuing decades, however, it became clear that what was at issue in this split of anthropology into physical and cultural anthropology was a fundamental debate over the constitution of the human sciences and their proper relationship to society. What methods were most appropriate for producing scientific knowledge about human beings? The experimental methods of the natural sciences, methods rooted in an evolutionary paradigm and used by physical anthropologists? Ethnographic methods rooted in the interpretive and qualitative approaches of the social sciences and used by cultural anthropologists? What, in any case, was the relationship between these methods and the legacies of race and colonialism?

These questions eventually proved so fractious that some departments of anthropology in the United States split into two separate institutional entities. For example, in 1998 the then fifty-year-old Department of Anthropology at Stanford University voted to divide into the Department of Anthropological Sciences and the Department of Cultural and Social Anthropology. The then acting chair of the Cultural and Social Anthropology Department, Sylvia Yanagisako, commented in an interview that "the intellectual issue for anthropologists involved 'a basic disagreement over what constitutes science' " (O'Toole 1998). Many other departments of anthropology in the United States exist with a de facto split in training programs, research schools, and power sharing.

Connected to these debates about methods were debates about the appropriate object of anthropological research. Beginning in the 1960s many anthropologists began to question the conventional idea within anthropology that "we" (i.e., "civilized people") could understand "ourselves" better by looking at "ourselves" in a "simpler" stage of development—a stage purportedly represented by indigenous groups cordoned off from modern social orders.[151] Rather than studying indigenous groups for what they could reveal about the "state of nature" in which humans evolved, physical anthropologists such as Frank Livingstone began to argue that indigenous groups should be studied for what they could reveal about processes of human adaptation to the natural environment (Interview T). In other words, anthropologists should not study indigenous groups because they were "primitive" and thus representative of the original human races as they existed prior to their "admixture" in modern societies, but rather because they often lived in unique natural environments (e.g., on top of a mountain or in a rain forest) and thus could provide insights into processes of human adaptation.

Social and cultural anthropologists provided a different critique of the anthropological tradition of studying so-called primitive cultures. This critique responded to the allegations by postcolonial scholars that efforts to study indigenous non-Western cultures merely resulted in the imposition of a Western objectifying gaze upon these cultures.[152] To oppose these objectifying tendencies, experimental strategies for writing ethnographies emerged that called into question the authorial voice of Western anthropologists studying non-Western cultures (Clifford and Marcus 1986, Marcus and Fischer 1986). Some social and cultural anthropologists also began to advocate for the study of Western as well as non-Western cultures (Stocking 1982 [1968], xii; Marcus 1998).

In short, by the time of the proposal of the Diversity Project in 1991, decades of self-critique had left anthropology both more divided and more self-reflective about its connections to racism and colonialism and its status as science. It is in this broader context that the Diversity Project organizers' effort to enroll anthropologists must be understood. As we will see below, the Project's difficulties would deepen as its proponents failed to recognize their entanglement in unresolved disciplinary questions within anthropology about how to study and know the human.

The Human Genome Diversity Project: Science or Racism?

As far as anyone recalls, questions about the Diversity Project's possible racist implications emerged from the floor after Cavalli-Sforza's 1992 AAPA plenary address. There the Howard University biological anthropologist Michael Blakey asked Cavalli-Sforza why he used the term 'ethnic groups' when referring to populations in Europe and 'tribes' when referring to groups in Africa. The use of the term tribe, Blakey implied, was the by-product of the colonial legacy of anthropology's past in which anthropologists distinguished between the so-called primitive people of Africa and the so-called civilized people of Europe (Interview T).

This implication that the Diversity Project was renewing anthropology's colonial past gained much greater visibility in April 1993 when the scientific journal *New Scientist* published an article on the Diversity Project. At the top of the second page the following assertion appeared in italics: "The project is 21st-century technology applied to 19th-century biology." The words were those of Alan Swedlund, a biological anthropologist who was at the time the chair of the anthropology department at the University of Massachusetts at Amherst. His implication was clear.

For many, the nineteenth century was the era of racism and colonialism; linking the Project to this time period was tantamount to linking it to the legacy of racism and discredited science. The *New Scientist* article went on to make Swedlund's critique explicit and to characterize it as central to a "heated debate about genetics, race and human welfare" that had pitted geneticists and anthropologists against each other in the debate about the Human Genome Diversity Project (Lewin 1993, 25).

Swedlund's quote in the *New Scientist* served to heighten the already-existing tensions between proponents of the Project and some critical physical anthropologists—including Blakey and Swedlund.[153] Cavalli-Sforza responded by charging Swedlund with unfairly maligning the Diversity Project (Interview T). Further, he threatened to withdraw from an upcoming November 1993 Wenner-Gren conference on the Diversity Project if Swedlund attended as planned.[154] This second Wenner-Gren conference had been organized by the University of Florida social anthropologist John Moore as a way to address anthropological concerns about the Diversity Project, and to build support for the Project within anthropology.[155] Moore believed that the Diversity Project had the potential to unify anthropology, as it would require the expertise of all of anthropology's four fields: physical anthropology, cultural anthropology, archaeology, and linguistic anthropology (Moore 1993).[156] In late summer of 1993, however, it was not clear that the Project's main scientific leader, Cavalli-Sforza, retained an interest in working with anthropologists, no matter what kind.

Cavalli-Sforza's response was consistent with his lifelong reputation as an anti-racist. In the 1970s, Cavalli-Sforza had on many occasions debated William Shockley, a Berkeley physicist who argued that women from "inferior races" should be sterilized (Cavalli-Sforza 1995). Through these and other responses to the race and IQ debates Cavalli-Sforza had come to think of himself as someone who debunked theories that used biology and genetics to justify racism. Given this personal history, Swedlund's accusation understandably upset him.

However, when understood within the broader history of attempts by human scientists to study indigenous populations, Swedlund's critique also makes sense. It is this history that Swedlund presented during his talk at the Wenner-Gren conference in November.[157] Studies of population history conducted before 1950, he told conference participants, had recognized "complexity in Europe and the difficulty in assigning race" but continued to regard "non-European" regions as still maintaining their " 'traditional' patterns and tribal structures." These anthropologists of the past used the categories ethnic group and tribe to mark a distinction between the purportedly impure mixed Europeans and the relatively pure non-Europeans. This distinction, he explained, derived not from nature, but from the subject-position of the researchers in Europe. As other anthropologists had noted, Europeans appeared more diverse because of the "significantly larger amounts of data available to European anthropologists on Europeans, as well as a greater understanding of the history of the region" (Swedlund 1993, 8–9). Ironically, it was exactly this gap in data collection that organizers of the Diversity Project hoped their Project would serve to correct. However, the way Cavalli-Sforza used terms such as tribe sparked fears that the Project might simultaneously promote anti-racism *and* racism.

Swedlund went on to argue that the view that tribes fell into discrete natural and social units served the interests of European states that needed to impose order to rule their colonial acquisitions. Citing the late Columbia University social anthropologist Morton Fried's 1975 study, *The Notion of Tribe*, Swedlund argued that tribes are the "social constructions of states that were superimposed by colonials upon non-European populations that were organized in all sorts of varying ways" (Swedlund 1993, 9; Fried 1975).

In short, Swedlund invoked well-established critiques of categories used in the colonial era: these categories did not reflect nature so much as they reflected the position and interests of the investigating subject—in this case, European researchers and colonial powers. By using the term tribe, as well as by viewing populations in Africa as divided into discrete social and biological units, Cavalli-Sforza, in Swedlund's view, fell into the trap of colonial science practices. He concluded by asking what evolution might look like if a group of Bantu scholars, and not the European-born Cavalli-Sforza, were to address questions of human diversity and evolution: "Would European genetic variation be viewed as more complex? How would biological modernity be defined?" (Swedlund 1993, 12).

Following this Wenner-Gren conference, deliberations about the Diversity Project's links to the legacy of racism and colonialism continued. However, rather than expanding and deepening deliberations about the Project, these discussions returned to old themes of expertise in a manner that served to narrow rather than broaden the debate. In particular, advocates of the Project once again asserted that problems would be solved by enrolling legitimate anthropological expertise.

This argument would be advanced by John Moore. In an article written shortly after the November 1993 Wenner-Gren meeting and entitled, "Is the Human Genome Diversity Project a Racist Enterprise?" Moore evaluated what he described as the "racist potential" of the Diversity Project (Moore 1996).[158] Drawing upon the authority of the recognized anti-racist physical anthropologist Ashley Montagu, Moore defined racism as the mixing of the following three "ingredients": the allegation that the human species is divided into biologically separate groups; the belief that individuals differ in immutable genetic traits; the conclusion that groups differ in genetic traits, and that these differences are meaningful in society and culture.[159] All three ingredients, Moore argued, had to be present for scientific racism to result. However, he maintained, as Montagu had warned forty-five years before, that the presence of even one ingredient was enough to create the potential for racism (ibid., 219).

On the basis of these definitions Moore concluded that the Diversity Project had the potential for racism. Through its use of migration theory the Project indeed did incorporate one ingredient of racism: a belief in the existence of separate biological groups. Thus, despite the fact that the Project's organizers were "certainly not racists, as their biographies and bibliographies clearly show," the Project nonetheless created what Moore considered "friendly" conditions for "racist interpretations" (Moore 1996, 228).[160] For Moore, the question then became: what could be done to prevent this potential from actualizing? His answer: the Project needed to adopt a "neutral and objective method" (ibid., 224). According to Moore, judgments about human character that led to racism were created by "dark forces" in society, and not discovered in nature by scientists.[161] To counter these distorting forces, organizers would have to conduct the Diversity Project in a "scientific manner" (ibid., 228).

But what would constitute a "neutral and objective method"? In answering this question, Moore returned to the theme of expertise. In particular, he echoed anthropologist contributors to the 1950 Cold Spring Harbor symposium, as well as NSF's Mark Weiss, in arguing that a rigorous scientific study of human genetic diversity could only come from incorporating anthropological expertise. Referring to one of Cavalli-Sforza's colleagues, the geneticist Alberto Piazza, Moore wrote: "With all due respect to Prof. Piazza's distinguished career as a medical geneticist, I must remark that he is not eminently qualified to comment on matters concerning the Neanderthal displacement or the origins of Paleolithic cave art," (1996, 223). These and other matters, Moore argued (and other Project organizers had already ceded), required the expertise of anthropologists.[162] Without them, he contended, the Project would at best do poor science, and at worst re-ignite scientific racism.

Importantly, Moore did not question the importance of genetic research. To the contrary, he agreed with Cavalli-Sforza that genetic analysis would lead the way to unraveling the mysteries of human diversity: "Quite correctly, I believe, he [Cavalli-Sforza] has taken the position that it is genetic analysis that will take the lead and provide the central framework for understanding distributions and changes in languages and culture on a continental and global scale" (Moore 1995, 531). His only point of difference with Cavalli-Sforza would be his belief that genetic analysis should be joined with proper ethnographic methods. Yet incorporating proper ethnographic methods and anthropological expertise is what the organizers of the Diversity Project believed they had been doing since responding to Mark Weiss's letter in *Science* and adding Kenneth Weiss to the organizing committee in late 1991.

Conclusion

At one level, organizers of the Diversity Project and the anthropologists who critiqued it shared the same view: human diversity was correlated with culture; thus, to study diversity it would be necessary to have cultural as well as biological expertise. Further, both critics and supporters agreed that for the Project to be legitimate science and not fall prey to racist practices, the geneticists who first proposed it would have to incorporate the expertise of anthropologists, the recognized experts on culture. However, as this chapter has demonstrated, disagreements arose about the nature of that expertise. Anthropological expertise, as both sides of the debate had difficulty recognizing, did not exist in an already-constituted form that organizers could merely add to the Project. Rather, this expertise would have to be produced through a set of complex negotiations that raised both conceptual and political questions about the structure of human societies and the nature of differences among them.

At stake in these negotiations were deeply embedded and contested questions in the human sciences about the nature of the human and who could speak for humanness. In particular, the effort to include physical anthropologists evoked long-standing debates about who had the knowledge and skills needed to study human diversity. In the 1960s, social and cultural anthropologists had served as assistants helping geneticists with their first expeditions to collect indigenous DNA. In the 1990s, however, physical anthropologists rejected this same role. Anthropological expertise, they argued, could not just be annexed onto an already formulated project; rather, it needed to be integral at every phase of design and implementation.

Yet even this stance proved problematic. Although all agreed upon the importance of something called anthropological expertise, disagreements emerged about what constituted this expertise. Although nominally all believed that the Project had to integrate a study of biology with a study of culture, differences emerged over the definition of culture and what it would mean to integrate biological and cultural approaches.

These questions about the constitution of expertise in anthropology and in the human sciences more broadly could not be separated from questions about the constitution of authority in society. Anthropologists had long faced questions about the role that their research on human populations in Africa and other parts of the colonized world had played in facilitating racism and colonialism. As Diversity Project organizers proposed to sample and study many of these same populations, their research inevitably raised questions about the place of the initiative in the broader histories of colonialism and racism, and even questions about who could define racism and colonialism in an era marked by the ascendance of expert genetic definitions of human differences.

The difficulty that both supporters and critics of the Project faced in acknowledging these historically entrenched questions prevented a needed deepening of the discussions about the Diversity Project. The assumption that expertise existed and could simply be added to (or included in) the Project impeded interrogations of the ways in which the Diversity Project raised questions about the very constitution of the goals and methods of the human sciences, and the relationship of these goals and methods to social agendas. Instead of opening up the Diversity Project to broader deliberations about the purposes of genetic sampling, supporters and critics alike merely asserted that already available expertise would help both to produce knowledge about human diversity, and to avoid the errors of the colonial past. The problems with this technical and insufficiently self-reflective approach would only become more evident as the debate about the Diversity Project spilled outside of the realm of scientific debate in the spring and summer of 1993.

Chapter 5

Group Consent and the Informed, Volitional Subject

In the midst of efforts to respond to critical physical anthropologists in the spring and summer of 1993, a second wave of criticism rocked the Diversity Project. At this time, organizers began to hear from indigenous rights organizations and other advocates for indigenous groups. In May, the Rural Advancement Foundation International (RAFI), an activist organization committed to policy advocacy on issues related to biodiversity and intellectual property rights, accused the Project of threatening the livelihood and autonomy of indigenous groups (Rural Advancement Foundation International 1993). In the same month, RAFI alerted indigenous rights organizations that Project organizers had prepared a preliminary Hit List of indigenous populations to sample. Shortly thereafter, in early July, the Third World Network called for an immediate halt to the Diversity Project (Native-L 1993a). Tensions escalated in the remaining months of 1993, culminating in December when the World Council of Indigenous Peoples dubbed the Project the "Vampire Project" (Indigenous Peoples Council on Biocolonialism 1998).

To address this new wave of concerns, Project organizers began to shift their energies from crafting what they understood to be scientific practices to creating what they believed would be ethical ones. In particular, organizers sought to create procedures that would ensure that only informed subjects who had voluntarily given their consent would participate in the Project.[163] This chapter describes why these efforts failed to assuage RAFI and indigenous rights organizations. By assuming the prior existence of a domain called science and a domain called ethics, and by treating each as a separate body of expertise, these efforts did not bring into focus the inextricable links between natural and moral order, and the role the Project would play in constructing each. Given that the bits of order at stake raised fundamental questions about the existence and status of groups in nature, and who could speak for these collectivities in society, not surprisingly these efforts proved controversial.

To illustrate this, the chapter critically examines two episodes in which Project organizers attempted to create informed subjects who might voluntarily consent to participate in the Project. The first episode centers on organizers' efforts to dispel the concerns of RAFI and members of indigenous rights groups by disseminating what they called objective information about the Project. Organizers believed that critics had merely misunderstood the status and goals of the initiative, and that they would have no reason to oppose it if only they correctly understood it. The chapter documents the debate that ensued over what could count as a correct understanding of the Diversity Project. Some indigenous rights leaders argued that the organizers' version of objective information obscured from view the conditions that made sampling in indigenous communities possible, and in so doing created a self-serving view of the Project that ignored the political and historical realities of colonialism and North/South relations. Diversity Project organizers responded that these accusations were themselves self-serving, acting to advance the interests of indigenous rights leaders. The chapter demonstrates that this asymmetrical analysis—one that attributed political content to the discourses of their critics, but not to their own—did not provide the framework needed to adequately clarify how the Project's research design and intent might be tied to issues of patenting, North/South relations, and power dynamics between researchers and research subjects.

These entanglements would be graphically illustrated by the second episode: the effort to construct group consent. Shortly following the second wave of critiques generated by advocates for indigenous groups, the North American Regional Committee (NAmC) of the Diversity Project formed an ethics subcommittee and began to draft what became known as the Model Ethical Protocol. Group consent emerged as one central innovation of this Protocol, and went on to become one of the NAmC's most celebrated responses to its critics. Expanding the standard Western biomedical practice of obtaining consent from individuals, group consent entailed obtaining consent from groups. The NAmC hoped that this additional form of consent would respond to the concerns of critics who feared that populations would be picked by researchers, and would not themselves choose to participate in the Project. Further, they argued that group consent represented an innovative cutting-edge contribution to the field of biomedical ethics. Indeed, in her statement to the Committee of the National Research Council that reviewed the Diversity Project in 1996, Mary-Claire King explained that the Project planned to "break new ground in recognizing the autonomy of communities," and hoped that "this standard [would] be applied beyond the HGDP to biomedical research generally" (King 1996).

However, far from the ethical gold standard that organizers hoped it would become, group consent only raised new worries. As illustrated in the pages that follow, this proposed ethical practice incorporated a number of assumptions: groups exist; groups are the appropriate unit of consent; groups can be defined for the purposes of consent; researchers' sampling criteria can be used to define groups; an authoritative voice exists for groups that can give consent. These assumptions—far from being neutral—removed from view questions about the role the Project might play in constituting human groups both in nature and in society. By assuming groups existed and could be defined by researchers, the proposed ethical innovation in effect closed off prior debates about the existence of human groups in nature. Further, by assuming that biological and cultural experts could define groups, group consent also sidestepped critical questions about the role the Project's research and ethical practices would play in defining groupness in society (Rabinow 1996). Because it did not provide the tools needed to bring into focus these critical issues, critics feared group consent might inadvertently grant scientific validity to old paternalistic and colonial notions of race and indigeneity.

From Scientific Objects to Ethical Subjects

As previous chapters have documented, up until and through the Penn State anthropology Workshop, the debate about the Diversity Project had been confined to one among scientists who claimed expertise in the area of human diversity. Thus, not surprisingly, representations and discussions of the Project employed an almost exclusively scientific idiom. Organizers and documents described those it wished to sample in terms of their genes, alleles, and membership in categories meaningful to geneticists and anthropologists. Debates focused on how these objects should be defined and sampled. To resolve the debates, all those involved, including critical physical anthropologists, agreed that proper scientific expertise would be required. They disagreed only on the nature of this expertise. As discussed in the last chapter, concerns focused on problems with what were viewed as the Project's proposed scientific objectives (i.e., sampling "indigenous populations," etc.), problems that critics believed derived from the use of categories defined using cultural and social, and not scientific, criteria.

During these first months of discussion, a consideration of the Project's entanglement with what are conventionally thought of as political and ethical questions remained largely absent. There were exceptions. For example, as noted in chapter 3, a brief passage in the original call for the Project in *Genomics* spoke of "peoples historically vulnerable to exploitation by outsiders" and the special problems of collecting samples from these groups. Additionally, the report from the first Diversity Project planning workshop concluded:

> It is the policy of the Human Genome Diversity project that all materials and data that are not confidential should be available to all qualified investigators. Ethical, legal, and human-rights issues connected with the project will be discussed in subsequent workshops. (Human Genome Diversity Project 1992a, 8)

And indeed, before the fall of 1992 Anthropology Workshop, these "ethical, legal and human-rights issues" did not form a focal point of discussion.

All this began to change, however, in the spring of 1993. At this time the Diversity Project began to receive the attention of critics of science, members of indigenous rights groups, and other advocates for indigenous people. The first sign of critical attention came from the National Institutes of Health. A couple of months before the Diversity Project's third planning workshop, program directors and other officials at the NIH (in particular, at the National Center for Human Genome Research) began to express concerns about the proposed sampling initiative.[164] Worried that it might turn into an ethical and political liability, the NIH urged Diversity Project organizers to devote a whole day of their third planning workshop, which was to be held on its Bethesda campus, to ethics (Interview E).[165] Diversity Project organizers promptly responded, and on February 17 convened one day of roundtable discussions on ethics and human rights at which collection issues, potential commercial implications of the Project, and the possibility that the Project would promote racism were discussed.[166]

The add-on of this day cast ethics as an instrument that could be fitted onto an already-formed research program, not as something integral to the conceptualization of research, as a co-production framework would dictate. Further, the workshop report represents ethical issues as manageable. It states: "There is no reason to believe that ethical concerns raised by the Project are insurmountable" (Human Genome Diversity Project 1993, 2). Finally, the value of the Project was undisputed, and the role of ethics was cast in instrumental terms:

> The value of this research and the urgency caused by the continuing disappearance of isolated human populations makes the ethical con-

> cerns all the more important. If the Project does not proceed carefully and properly, it could spoil the last good opportunity to obtain some of this data. (ibid., 3)[167]

In assertions such as these, the reason to do the ethics right was not to address the legitimate concerns of subjects, but rather to protect the Project from unwarranted criticism that could harm its future.

Only a few months after the Bethesda meeting, though, it became evident that this instrumental approach to ethics, and the representation of the Diversity Project as ethically manageable, might not hold. It was at this time that Project leaders first heard from the Rural Advancement Foundation International. RAFI was an Ottawa-based international research organization devoted to policy advocacy on issues relating to biodiversity and intellectual property rights. Although unknown to most Diversity Project organizers, since the 1970s this organization had been at the forefront of a movement that connected the identity and survival of indigenous populations and rural farmers in the global South to biological diversity. RAFI, along with others who took their case to the United Nations Food and Agriculture Organization (FAO) in the late 1970s, argued that Northern governments and corporations should not be allowed to patent germplasm from the South. The late 1970s and early 1980s witnessed heated debates on this topic both at the FAO and in the assemblies of national governments. Resulting national laws, U.N. commissions and international legal accords endorsed the principle that genetic diversity (here defined as seeds) was essential to the livelihood and survival of the farmers and indigenous populations of the South (Mooney 1996, 21–30).

Upon reading about the Diversity Project in *Science* in late fall of 1992, members of RAFI worried that the Project raised the same issues as biological diversity, but now in connection with human integrity and autonomy. RAFI focused its concerns in a letter sent to Diversity Project organizers on 6 April 1993. When Diversity Project organizers did not respond to this letter, RAFI issued a communiqué on May 15 in which it made public its worries about the Project. The communiqué reported that the Diversity Project represented the same if not a greater threat to the livelihood and autonomy of indigenous groups as plant prospecting had in the 1970s (Rural Advancement Foundation International 1993). Specifically, RAFI argued that the Project threatened to exploit "poor people whose survival is in question" by patenting products derived from their genes; diverting money for basic services, such as access to clean water and vaccinations, to the Diversity Project; transferring human genetic resources to gene banks in industrialized countries, such as the United States; and making it "theoretically possible for unscrupulous parties to devise cheap and targeted weapons effective against specific human communities" (ibid., 4).

At the same time, RAFI re-released the list of populations that had been prepared at the Penn State anthropology workshop under the title Hit List and an epigram that called attention to the list's use of "inconsistent and/or antiquated geographic references and several impolite and/or antiquated names for indigenous peoples" (Rural Advancement Foundation International 1993b).[168] RAFI sent both documents to several international indigenous rights organizations.

RAFI's objections were the first of many that would highlight the ways in which Diversity Project organizers were working on a highly contested terrain that had already been defined by values, practices, and ideas other than those of the scientists who to date had been the primary participants in discussions about the Project. In particular, the Project was preceded by decades of debate and struggle in which the identities of human groups had been articulated to genetic diversity. These identities were rich with multiple meanings, power relations and contentious fractures. To move forward, Diversity Project organizers would have to order and stabilize this already fractured terrain. This task would prove difficult to manage.

THE INDIGENOUS RIGHTS MOVEMENT

The entanglement of the Diversity Project in these broader political struggles over cultural identity became even more evident in the summer of 1993. At this time criticism of the Diversity Project spread beyond RAFI to indigenous rights organizations and networks. As this happened, fundamental political questions about the role the proposed Diversity Project would play in constructing the identity of indigenous peoples moved to the fore.

As many organizers of the Diversity Project would begin to learn, subjective modes of determining identity, such as self-identification, had gained authority in a variety of national and international institutionalized contexts in preceding decades.[169] During this time, human identity had become not just an object of scientific knowledge, but also a resource that people strategically used in struggles over rights, equity, and power (Cohen 1985). Indeed, gaining the power to define one's own future in keeping with one's self-identification formed a central goal of anti-colonial movements during the 1960s and 1970s (Warren 1998).

One strand of this broader history is that of the indigenous rights movement—by the 1990s, a two-decades-long effort to build national and international support for the rights of indigenous peoples. This effort derived its principles in part from the attempt to define and establish basic human rights in international law following the Second World War. Based on the principles of an emerging universal humanism, many hoped that the discourse of human rights would act as an antidote to the differentiating logics that had been used to justify the slaughter of millions during World War II. Instead of a world divided into superior and inferior individuals and groups, universal humanism imagined the "united family of man" (Haraway 1989, 198). This doctrine became embodied in institutions such as the United Nations, and in official documents, such as the Universal Declaration of Human Rights.

Nonetheless, governments continued to treat some humans, in particular members of indigenous groups, as less than human; acts of genocide did not cease with the end of the Second World War. In the early 1970s, soon after a massacre in Latin America gained worldwide attention, international organizations, scholars, and activists began to work to mobilize indigenous peoples internationally to respond to this continued oppression and to secure their basic human rights (Interview O). Several changes in international law quickly followed. In 1971 the United Nations decided that indigenous issues were rightfully deliberated at international institutions like the World Court and the United Nations, and were not the exclusive province of member states (Warren 1999, 6). In 1974 the International Indian Treaty Council (IITC), the international diplomatic arm of the American Indian Movement (AIM), became the first indigenous organization to receive United Nations Type II (Consultative) nongovernmental organization (NGO) status (Robbins 1992, 106). In 1975, the World Council of Indigenous Peoples (WCIP), a separate international body for indigenous issues, was formed (Sanders 1980).

By 1977, international governing bodies recognized the existence of an indigenous rights movement.[170] In that year 130 people representing Iroquois from the United States, Guaimi from Panama, Aymara from Bolivia, and dozens more indigenous groups from around the world went to a Non-Governmental Organization meeting at the United Nations. At this meeting they demanded to be recognized as real people living in a real world different from that of Western Civilization. As some who participated in this historic gathering later recounted:

> All week there had been pressure felt. At first slowly, but then fairly rapidly the word had spread that in this conference no one was holding back—finally, thoroughly and uncompromisingly that Indian peoples of the American continents had not died, were not about to die, that they may now be cultures within cultures and Nations within Nations and that their oppression may have been long and arduous—the cruelest maybe in recorded history—but that if some things had been lost, nothing had been given up, nothing. (Akwasasne Notes 1995 [1978], 37)

To support their claims for distinct recognition, delegates to the U.N. meeting cited governments, religions, cultures, and economies that did not contradict nature, but rather fit with it. This harmony with nature, they argued, defined them as a people (indeed, they referred to themselves as Natural World Peoples). As a self-proclaimed real people living outside of Western Civilization and in nature, they now demanded full recognition by the United Nations.[171]

Given this history, it is not surprising that the proposal for a Diversity Project alarmed many indigenous rights organizations. In almost all of the documents that announced the Project, as well as workshop reports, indigenous groups were described as "vanishing" or "disappearing," groups of historical interest that needed to be "preserved" for study before they "lost" their "identity" (Cavalli-Sforza et al., 1991, 490; Human Genome Diversity Project 1992b, 5). Yet this was exactly the representation that indigenous groups themselves opposed.[172] Not only were they not disappearing, they were here to stay and demanding protection of their rights as well as representation in international governing bodies, just as any other free peoples would.

In an age of electronic communication, it took very little time for a group to broadcast this problematic connection between the indigenous rights movement and the Diversity Project (Lock 1994). On 8 July 1993, the Third World Network posted the following message on Native-L, an aboriginal, First Peoples news net:

> URGENT! URGENT! URGENT!
>
> ***************************
>
> CALL FOR A CAMPAIGN AGAINST THE HUMAN GENOME DIVERSITY PROJECT

The notice that followed argued that the Diversity Project put an interest in historical curiosity above an interest in the future well-being of indigenous groups. The Project's focus on preservation, and its use of terms like "Isolates of Historical Interest," demonstrated a preoccupation with the past rather than the future (Native-L 1993a).[173] This focus represented indigenous groups as objects of historical interest that were about to go extinct as opposed to "fully human communities with full human rights"— a stance they described as particularly egregious in 1993, the U.N.-decreed year of indigenous peoples.[174] In short, by redescribing indigenous groups as "vanishing," the Project threatened to once again reduce them to objects of mere scientific interest. Authors of the Third World Network message worried that this would enable both their debasement and their commodification.[175]

To place these detrimental developments in context, authors of the message connected the Diversity Project to the emergence of "new biotechnologies" and the formation of the Human Genome Organization, and implied that the Project arose out of Western economic interests that sought to transform the genetic differences of indigenous peoples into dollars (a process that critics would soon label biocolonialism). The Network called upon "all groups and individuals concerned with indigenous peoples' rights to mobilize public opinion against the case of human communities as material for scientific experimentation and patenting" (Native-L 1993a).

Not only did advocates for indigenous peoples fear that the Diversity Project would rework the identity of indigenous peoples in a disempowering manner, they also objected to Diversity Project organizers' constructions of who could speak for them. This issue first emerged in the summer of 1993 in an exchange between RAFI's executive director Pat Mooney, and Henry T. Greely, Stanford University law professor and organizer of the Diversity Project's February 1993 workshop on ethics.[176] Mooney pointed out that Greely had invited Jason Clay of Cultural Survival Enterprises Inc. and Walt Reid of World Resources Institute to the February 1993 Bethesda ethics workshop, but no members of indigenous groups. Mooney criticized this choice, and argued that it was "no more appropriate to have Cultural Survival speak on behalf of indigenous peoples than for RAFI." Further, Mooney challenged the Project's representation of indigenous peoples as just cultural groups that lived in villages, groups that had no "organizations of their own" that could speak for them. "This is," he corrected, "simply not the case. . . . Indigenous peoples speak for themselves" (Mooney 1993).

Within months of this exchange, members of indigenous rights organizations issued similar critiques. For example, on 20 October 1993, Nilo Cayuqueo, then co-director of the South and Meso American Indian Information Center (SAIIC), posted an alert on the Native-L that called for an immediate halt to the Diversity Project until "all parties had been properly consulted." As the message explained: "The fact is that from beginning to end, the very people from whom the samples are being taken are not being consulted during any stage of the process" (Native-L 1993b). To correct this, among other things, the alert called for a meeting between the World Council of Indigenous Peoples (WCIP) and organizers of the Diversity Project.[177]

A PROBLEM OF INFORMATION?

Diversity Project organizers proved open to some of these criticisms and suggested remedies. In particular, they expressed a desire both to consult with indigenous groups and to advance what they viewed to be their interests.[178] As Greely wrote in a letter to Pat Mooney on June 8:

> I want to stress again that the Human Genome Diversity ("HGD") Committee wants to work with indigenous peoples and their organiza-

> tions to ensure that this kind of research, which will happen with or without the HGD Project, is done in ways and in circumstances that advance the rights and interests of indigenous peoples and not undercut them. The HGD Committee thinks that the Project will help indigenous peoples, not harm them, and we will work hard to make sure that this is the case. (Greely 1993b)

Greely went on to take the blame (along with a Boston snowstorm that prevented Clay from attending) for the lack of "voices sensitive to the interests of indigenous peoples." However, he argued that the absence was one of "ignorance, not intent." He concluded by telling Mooney that he hoped to "shamelessly" use him and RAFI "as a way to become better informed about the organizations and people who represent the interest of indigenous people" (ibid.). Greely repeated this same point in his letter to Mooney of July 8: "The HGD Project organizing committee sees the participation of indigenous peoples as essential to the Project's success and it supports their interests" (ibid., 1993c). Finally, as further evidence of the Project's commitment to consulting with indigenous groups, in December 1993 a representative of the Project traveled to Guatemala to present the case of the Diversity Project to the WCIP.

Project organizers proved less open to the substance of the critical messages posted by the Third World Network and SAIIC on the Native-L. These messages, they argued, had been based on "factually incorrect" information and were not justified. As Cavalli-Sforza and Greely stated in their e-mail response to the Third World network alert:

> Today we read with great concern a message on this network seeking to enlist support to stop the Human Genome Diversity (HGD) Project. That message conveyed *factually incorrect and grossly misleading information* about the proposed Project and, more importantly, was deeply wrong in its stance toward that proposed Project [italics added] (Greely and Cavalli-Sforza 1993).

To correct their critics, Greely and Cavalli-Sforza offered the following information:

- the Diversity Project did not yet exist; it was still in the planning stage and so not yet collecting, preserving or analyzing any samples;
- the Diversity Project sought to increase the "body of knowledge about all the earth's people," not just the "citizens of the United States and Western Europe," the subjects of existing human genetic research;
- analysis of human genetic variation could lead to the answers to "many fascinating questions" about human evolution, history, linguistics, adaptation and even disease;

- by including indigenous peoples in this research, the Project did not intend to exploit them, but rather to help them by extending the insights and benefits of genetic science to the developing world;
- the Project would only "succeed" if it had the "participation of the sampled populations and if it fully respect[ed] their interests and autonomy";
- human tissue samples would be taken only with the "full informed consent of the donors";
- samples would be stored both in the United States and in regional repositories around the world;
- the Project's goal was to conduct this research in a visible, accountable and coordinated way that met with the "highest standards" of science and ethics;
- the Project was not and would not become a commercial venture; the chance that the research would lead to the development of commercially valuable products was "very remote." (Greely and Cavalli-Sforza 1993)

All of this "accurate information," organizers argued, contradicted accusations made by the Third World Network that the Diversity Project had already begun, that it would economically exploit indigenous populations, and that it would be a museum preservation project.

In future postings to Native-L, Greely continued to argue that his messages were intended to provide "accurate information" (Greely 1993e). This tactic soon generated debate as critics began to call into question what could count as accurate information in the Diversity Project case. Perhaps the most heated debate about this issue arose when in the summer of 1993 RAFI questioned the Diversity Project's role in the patenting of indigenous DNA. This episode brought into relief Project organizers' difficulty in recognizing the potential political and economic content of their Project, a difficulty that would continue to trouble their attempts to create a social and moral order that could support their endeavor.

COLONIALISM OR ANTI-COLONIALISM?

The issue of patenting arose very early on in the Diversity Project debates. As noted above, Project organizers devoted one session of the 1993 Bethesda ethics workshop to a discussion of the potential commercial applications of the Diversity Project. At this meeting, organizers denied that their Project had any interest in pursuing patents. They did not deny that patenting would be an important issue in human genetic diversity research (as pharmaceutical companies would undoubtedly pursue them), but they did deny that scientists involved in the Project had any interest in this dimension of the research. Further, heeding the advice of Dr. Walter Reid (a member of the World Resources Institute who had participated in the attempts to preserve worldwide biodiversity at the Biodiversity Convention in Rio de Janeiro in 1992), and the human geneticist Val Giddings (who had been on the United States's negotiating team for the Biodiversity Convention in Rio, and who would go on to become vice president of Food and Agriculture at the Biotechnology Industry Organization), organizers decided to make a "preemptive strike on the patent issue" by stipulating that no one could patent genes discovered using cell lines made for the Diversity Project (Human Genome Diversity Project 1993, 14; Greely 1993d).[179]

RAFI, through its executive director Pat Mooney, questioned this position. Mooney argued that this policy contained two loopholes that the organizers themselves had recognized at the Bethesda ELSI meeting: (1) one could stipulate that no patents could be sought for genes taken from the Project's samples, but a company interested in patenting such genes could go back to the country where the gene was found and make its own financial arrangements; and (2) the stipulation against the patenting of genes would not prevent the patenting of proteins produced by these genes (Human Genome Diversity Project 1993, 14; Mooney 1993b). These loopholes led Mooney to conclude that although the Project's decision to prohibit patenting of genes was "laudable," it would not keep human materials collected using U.S. government funds outside the patent system (Mooney 1993).

RAFI deemed the potential for the patenting of human materials a critical problem, and in the summer of 1993 began a public campaign on this issue. Using the American Type Culture Collection database, in late summer 1993 RAFI found a patent on cell lines of a "26-year-old Guaymi Indian Patient in Panama" who was infected with the HTLV-II virus (Mooney 1996, 123). Writing to Greely on June 30, Mooney cited this patent as evidence that patenting would be an issue for the Diversity Project. The Guaymi, after all, were among the populations listed by the Diversity Project in the Penn State anthropology workshop report (the very report RAFI had dubbed the Hit List).

Greely subsequently acted to oppose the Guaymi patent. As he explained in an e-mail message posted on Native-L on October 25:

> The Project most emphatically does not want to patent anything from these samples. Its organizers are not involved in this Project to make money, nor do we wish to repeat the sorry story of plant genetic diver-

> sity and the developing world. Although we think it is highly uncertain that any products of commercial value will arise from the Project, we are committed to ensuring that financial benefits from any such products flow back to the sampled populations. Implementing that commitment involves tricky questions of U.S. and international patent and contract law, but we will resolve them. In fact, I spent a good part of last week working (with the World Council on Indigenous Peoples and with Pat Mooney, Executive Director of RAFI, of which more later) to get the U.S. government to withdraw a patent application it had filed for a cell-line made from blood donated by a Native American from Panama (from the Guaymi people) for epidemiological research. (Greely 1993e)

Rather than serve to promote Western (in particular, American) imperialism through patenting indigenous DNA—as RAFI, and later SAIIC, suggested—Greely argued that the Project opposed patenting. He also claimed that the Project would oppose American imperialism through expanding the scope of human genomics to include all the world's people. The Human Genome Project, he suggested, was parochial in its ambitions:

> The Human Genome Project, a $175 million per year "joint venture" of the U.S. National Institutes of Health and the U.S. Department of Energy, aims to sequence entirely the 3 billion base pairs of "the human genome," but with its concentration on North America and Europe, it will in effect tell us everything about the genes of one French farmer and one little old lady from Philadelphia—but nothing about the rest of our species. (Greely 1993f)

By contrast, through expanding studies of human genomics to include analysis of populations from all over the world, Greely argued that the Diversity Project would help demonstrate "how closely related all humanity is — that we are, in literal fact, an extended family." Creating this "kind of understanding," he concluded, "should help all the world's human populations" (Greely 1993e).

These acts and claims, however, did not convince Mooney and RAFI. Patenting, they argued, remained a serious concern. Indeed, in a press release issued the day after Greely notified Native-L of the Project's actions to oppose the Guaymi patent, RAFI charged:

> Behind this [the attempt to patent the Guaymi cell line] lies the whole Human Genome Diversity Project of Europe and North America, which is planning to collect DNA samples from ten to fifteen thousand indigenous persons from more than 700 ethnic communities worldwide, at a cost of US $23–25 million over five years. (Rural Advancement Foundation International 1993c)

Not surprisingly, Greely was frustrated. He had been in direct contact with Mooney about the work he was doing to oppose the patent. "After those efforts," he wrote to Mooney on November 1, "I was offended that your press release implied the HGD Project *supported* the patent application." He concluded: "RAFI owes a duty—to those who read its press release and to the truth—to correct its misleading references to the HGD Project" (Greely 1993d).

Mooney responded the next day. His message was in no way conciliatory or apologetic. Instead, he stood by the claim in the news release that the Diversity Project was "behind" the Guaymi patent claim, arguing that it meant exactly what Greely had thought in his October 27 e-mail to Mooney—"that concern for this patent application is heightened by the pending HGDP Project where the danger exists that human cell lines collected through the project will be patentable." He promised that if a report should emerge that gave any other interpretation, he would clarify the statement directly with the media involved: "End of Point" (Mooney 1993c).

Second, Mooney rejected Greely's claim to have been a primary player in efforts to revoke the patent. He might have helped—and Mooney thanked him for that help—but the main work had been done by others: in particular, the President of the Guaymi General Congress Isidro Acosta and his colleague who went to Geneva in October to address the intergovernmental meetings of the Biodiversity Convention (the Convention's first meetings since the Earth Summit in Rio), and to meet with senior officials of the General Agreement on Trades and Tariffs (GATT) in the Intellectual Property Secretariat. Mooney also gave primary credit to the organizations that had supported the trip (the World Council of Churches, the Worldwide Fund for Nature, and Swissaid).

Mooney also argued that in opposing the patent the Project looked less like a protector of indigenous groups, and more like a self-interested organization. RAFI knew (from the transcript of the Bethesda ELSI meeting sent to them by Greely) that Project organizers had been told that patenting genes would be politically very sensitive in many countries, and to move forward they would need to make a convincing stand against patenting. Early on (in his letter of June 30), Mooney noted that if it appeared that the Project would benefit financially in any way, the credibility of the Project would be jeopardized. He later repeated:

> I have no doubt that you are personally opposed to the patent claim. I also have no doubt that the HGD project recognizes that the pursuit of such a claim would fundamentally jeopardize its ability to collect human cell lines overseas. I assume that both interests were at work, and I would expect that the National Institute of Health and the Center of Disease Control were anxious not to put the Project at risk. (Mooney 1993c)

In other words, the question of whose interest the Project represented when it opposed the Guaymi patent remained unsettled for RAFI. Organizers might sincerely want to stop the Guaymi patent, but what, RAFI asked, were their motives? To protect the rights of indigenous groups, or to promote the Diversity Project?

Finally, Mooney argued that the medical research establishment controlled funds that would likely support the Diversity Project, and so the Project would inevitably contribute to medical research, and in the current climate anything that contributed to medical research could be patented.

Very soon after Mooney raised these objections, SAIIC's Cayuqueo also called into question Greely's portrayal of the patenting issue. In particular Cayuqueo questioned Greely's description of the blood that led to the cell line as "donated":

> The notion that the Guaymi sister in Panama "donated her blood" indicates the narrow and paternalistic way in which the people involved regard the situation. It is still not clear how she was approached and why the President of the Guaymi General Congress, Isidro Acosta, was not involved from the onset. To imply that the woman would have willingly given her blood sample had she known the U.S. government was going to then apply for a patent on her cell-line is ludicrous. Your description does not reflect the reality of the situation, Mr. Greely, and this is very dangerous. (Cayuqueo 1993)

Further, Cayuqueo called into question the likelihood that indigenous populations of interest to the Diversity Project would ever find themselves in a situation where they would be "donating blood." Donation implied choice. But, he asked, "Given the history of the past 500 years, where and how does choice factor in?" (ibid.).

These disagreements over what constituted an accurate representation of the Project's role in the patenting of indigenous DNA continued to be a source of tension between Diversity Project organizers and advocates for indigenous groups. RAFI and indigenous rights organizations, such as SAIIC, continued to argue that given the broader political and economic context in which the Project would be unfolding, assertions that the Project could oppose the patenting of biological materials collected as part of the sampling initiative were naive at best, and self-interested at worst. For their part, Diversity Project organizers continued to discount charges that their Project would play any role in the growing commercialization of human genetic research. Claims to the contrary, they insisted, misrepresented the Project. They hoped that these misrepresentations could be overcome through communicating what they viewed to be accurate information. Further, they asserted that an understanding of the Project's true goals and intentions should lead indigenous peoples to support the Project. While wishing to communicate with representatives of indigenous peoples groups, organizers compromised the possibility of such dialogues by dismissing the concerns of the Third World Network and SAIIC, thus foreclosing important discussions of the ways in which the Project might become intertwined in complex issues of commercialization and indigenous rights.

Beyond asserting that indigenous groups simply had not been informed, organizers later argued that charges by these groups were "politically motivated." This, for example, would be the assessment of the Project organizer who represented the Diversity Project at the 1993 World Council of Indigenous People biannual meeting, held that year in Guatemala.[180] After his talk at the WCIP, the floor was opened up for questions. But there were very few questions. Instead, over the course of the next hour and a half the Project representative found himself the object of several denunciatory speeches. Among other things, WCIP delegates (most of whom were from the South) accused him in Spanish of being a CIA agent who had come "only to further the secret military interests of the U.S. government" (Interview 1999). All the criticism, he recalled, began with the "proposition that this is obviously just another piece of exploitation by Europeans." Such a position, he argued, precluded the possibility of rational dialogue: "You can't argue with that. You can't discuss with, you can't assuage that kind of concern. Everything you say then to point out the protections and the interests and issues that you are taking into account becomes a part of the oppressor's plot." This kind of criticism, he concluded, did not derive from facts, but rather from "political motives" (ibid.).[181]

On the one hand, the Project organizer's frustration with the denunciatory speeches is understandable. On the other hand, to dismiss them as politically motivated was to miss a chance to understand the Project's place in much deeper histories of colonialism and North/South relations. While recognizing the ways in which the Diversity Project raised fundamental questions about the definition and status of groups as biological entities, at events like the WCIP meeting in Guatemala, organizers failed to adequately recognize that their Project would also raise fundamental questions about the definition and status of groups in society, especially those occupying structurally disadvantaged positions. Instead of admitting that their science project might have political and economic content, organizers dismissed the critique of indigenous rights groups as political statements designed to advance their own agendas. This asymmetry in the Project organizers' analysis—attributing political content to the discourse of critics and not their own discourse—would continue to trouble organizers as they attempted to respond to the concerns of RAFI and indigenous rights groups. The debate over group consent illustrates these points.

CHAPTER 5

Group Consent: A Fundamental Debate about the Order of Nature and Society

Following criticism by RAFI, the Third World Network, SAIIC, and an increasing number of indigenous groups in the spring and summer of 1993, the newly formed North American Regional Committee appointed an ethics director, Henry T. Greely (the Stanford law professor who had organized the February 1993 Bethesda ethics workshop).[182] Shortly thereafter, Greely and Marcus Feldman, another member of the NAmC, applied for funds from the John D. and Catherine T. MacArthur Foundation to write what became known as the Model Ethical Protocol.

For members of the NAmC, these developments represented a shift in focus from science to ethics. Prior to the second wave of attacks on the Diversity Project, one goal of the NAmC had been to secure funds for writing a Model Protocol. This Protocol, as originally conceived, would guide scientists as they wrote grants to support the Human Genome Diversity Project pilot studies.[183] As Project documents indicate, members of the NAmC intended that it would address the following areas: criteria for selecting populations; sampling procedures; defining ancillary information; sample storage issues; access to data and samples; ethical issues. Although such a document would have addressed what they considered ethical issues, the main goal was to provide guidelines for what they understood to be the science (Interview P, Interview A4). Indeed, the very term "protocol," one organizer recalled, was a scientists' term.

The focus of this effort, however, changed in 1993. At this time, a Diversity Project organizer contacted the Special Projects office of the MacArthur Foundation to ask about possible support for the Project. The MacArthur Foundation, known to support work at the boundaries of disciplines, had in the past supported the work of a Diversity Project organizer and it had been thought that perhaps the Special Projects office might have an interest in supporting the pilot studies for the sampling initiative. The organizer soon received word back that the Special Programs Office did not fund science. However, it could fund efforts to build communication between the Project and Native American groups for the purposes of drafting ethical protocols (Interview B).[184] With the critiques of RAFI and indigenous rights organizations foremost in their minds, organizers realized that communication efforts with those outside the Project would be essential and in November 1993 submitted a proposal to the MacArthur Foundation titled, "Pilot Studies on the Human Genome Diversity Project." Among other things, the proposal asked for money to fund preparatory work for an ethics subcommittee, including the drafting of a "model protocol."

As Project documents indicate, in February 1994, a program officer at the MacArthur Foundation contacted Project organizers to inform them that the proposal did not adequately address the need to communicate with Native and indigenous communities about the Project. To remedy the problem, they recommended that organizers write an addendum that would outline a communications plan.[185] Organizers quickly provided such an addendum. In March 1994 the MacArthur Foundation agreed to fund their proposal.

In the coming year-and-a-half the NAmC would devote much of its energies to the drafting of what became known as the Model *Ethical* Protocol.[186] This redirection of energies toward the drafting of ethical protocols makes sense within a broader historical context. The Diversity Project was proposed in the American setting where in the late 1960s research ethics had become a major concern. At this time Henry Beecher, Dorr Professor in Anesthesia at Harvard Medical School, published an article in the *New England Journal of Medicine* documenting twenty-two cases in which investigators had endangered or even destroyed the health or life of their research subjects (Beecher 1966). Beecher's article and new concerns about impacts of scientific research (in particular, nuclear technologies and pesticides) sparked major changes in the regulation of human research (Rothman 1987). Focus shifted from the utilitarian value of biomedical research to the rights of research subjects.[187] The rise of the biotechnology industry during the late 1970s and 1980s also created new concerns about the commodification of knowledge and social order (Hilgartner 2002). With these changes, research ethics, once infrequently discussed, became the subject of numerous articles and federal regulations. Calls for formal review procedures proliferated, and in time a new profession and institutionalized space for decision-making known as biomedical ethics (or bioethics) came into being to meet the new demands for oversight and governance (Kleinman, Fox, and Brandt 1999). Diversity Project organizers' turn to ethics to provide oversight and regulation of their initiative fit within this longer historical trajectory.

Yet despite the devotion of significant time and energy to this widely accepted practice of designing ethical guidelines, the NAmC's efforts failed to address the concerns of their critics. Below I examine why, providing two interconnected explanations. First, the move to respond to critics by developing new ethical practices cordoned off the complex political questions about North/South relations and indigenous rights into the narrow technical domain of professionalized biomedical ethics. Second, Project organizers' understanding of ethics rested upon the incorrect assumption that the science of the Diversity Project could be separated from its ethics. It thus proved difficult for these organizers to see or accommodate the epistemological issues at stake in creating ethical practices. We turn now to the example of the group consent provision of the Model Ethical Protocol for a more detailed elaboration of these two points.

NAmC GROUP CONSENT

According to NAmC Project organizers, the logic of group consent was first convincingly articulated at the international planning meeting for the Diversity Project held in Sardinia in September 1993. At this meeting John Moore, organizer of the upcoming Wenner-Gren conference on the Diversity Project, explained how the American Indian tribes he worked with handled consent issues.[188] In a number of these tribes, Moore said, tribal elders met to make decisions for the whole group. As a result, he sought consent for his research from individuals as well as these groups.

Moore's informed-consent practices departed from standard codes of ethics for Western biomedical research. According to these codes, for research to be deemed ethical, informed consent must be obtained from the individual research subjects who are assumed to have personal needs, and preferences that are entitled to deference. The Belmont Report, the definitive American statement on informed consent, explains that this ethical practice follows from the conviction that "individuals should be treated as autonomous agents," a principle that is consistent with the liberal democratic tradition of protecting and preserving *individual* autonomy and rights.[189]

In its group consent provision, the Model Ethical Protocol adopts Moore's extension of informed consent from individuals to *groups*. Project organizers argued that this extension would ensure that groups had freely chosen, and had not been picked (as critics charged), to participate in Diversity Project research. Yet, although designed to quell controversy, group consent only generated new concerns. As we will see below, group consent was premised on assumptions that did not address questions about how groups are socially constituted and valued in liberal democracies, and defined by scientists for the purpose of ordering nature. Instead of addressing these fundamental questions, many argued that group consent would in effect produce groups that had questionable status in both society and nature. The particularities of these arguments provide a vivid demonstration of the ways in which natural and social order are crafted together.

GROUPS ARE THE UNIT OF CONSENT

Group consent, as conceived by the NAmC ethics subcommittee, was based on the following assumption: if groups (for example, "indigenous populations") are the subjects of research, then they also should be the subjects of consent.[190] As the Model Ethical Protocol explains:

> Although this requirement [group consent] goes beyond the strictures of existing law and ethical commentary, we believe it flows necessarily from the nature of the research, which is, by definition, research aimed at understanding *human populations* and not individuals [italics added]. (North American Regional Committee 1995, 3).

This is one of several passages in the Protocol that justifies group consent by appealing to the "population-based nature" of the research (ibid., 9).[191] Unlike in the earlier debates between Cavalli-Sforza and Wilson, in these passages the status of populations is no longer in question. Instead, they are assumed to be stable enough to serve as the consenting subject.

The Protocol adds a second reason for seeking group consent:

> [T]he effort to include samples from throughout the human species means that many of the populations sampled will not be part of the industrialized world, where genetic studies to date have concentrated. Many of the populations that might participate in the Project are politically or economically marginal in their countries. They have faced discrimination, oppression, and even genocide. Under such circumstances, it cannot be ethically appropriate to sample some members of a group when the group itself has not agreed to participate in the Project. Such methods would themselves be another form of attack on the autonomy of the population. (ibid., 8).

Drafters of the Model Ethical Protocol designed group consent with the hope that it would explicitly recognize indigenous groups' status as subjects and not just objects, and thereby produce a moral order that could support their project.

It is harder to see the role that group consent played in producing natural order. This becomes apparent, though, if one locates the discussion of groups in the Model Ethical Protocol in the larger context of Diversity Project organizers' debates about human groups. The Protocol states that groups are the objects of research. Indeed, the assertion that groups are the objects of research becomes the very basis upon which groups are positioned as the subjects consenting to research. However, this understanding of groups—that they would be the objects of research in any study of human genetic diversity—differed from the understandings that shaped the early Diversity Project debates about the design of a sampling strategy. In these debates, the status of human groups was less clear. As we saw in chapter 4, some did indeed argue that groups (for example, populations) structured the genetic diversity of the human species. Others, however, argued that this might not be the case in all parts of the world. Still others argued that human groups might structure the genetic diversity found in nature, but that to determine the boundaries of these groups, one would have to sample individuals on a grid. But in the text of the Model Ethical Protocol these contested questions about the status of natural groups were shut down. Instead, the proposed group-consent provision assumed that natural order could be usefully represented by groups, that Diversity Project participants would sample and study groups, and that moral considerations would be served by giving informed-consent rights to groups.

DO GROUPS EXIST?

Not all Diversity Project organizers supported this construction of groups as natural and morally recognizable entities. European organizers, for example, argued that social and cultural groups would not necessarily be the appropriate objects of research. Indeed, hearkening back to the ontological debates about the status of groups in which Wilson questioned the existence of culturally and linguistically defined biological groups, some European organizers asked whether social and/or cultural groups even exist. As one organizer recalled:

> [T]he Europeans were against this [group consent] because they don't need it and they are afraid that it may generate enormous problems for them.
>
> JR: To have to identify these groups?
>
> Yeah, there are no such groups. So they'll have to create them in order to talk to them. (Interview E)

In Europe, this organizer noted, many potential studies of human genetic diversity would not involve groups.[192] Thus, if groups were going to give their consent for these studies, the researchers would first have to create them. In other words, groups would not preexist the research design.

The assumption that groups could be the consenting subjects as well as the objects of research brings to light the entangled character of natural and moral order. In order to build a scientific project to study an aspect of natural order, organizers recognized that they had to build a moral order that could accommodate their project. To do this they attempted to extend informed consent practices. Groups as well as individuals, they argued, would have informed consent rights. Yet, as they attempted to construct this moral foundation, organizers ended up revising their claims about the natural order—in particular, the claim that groups exist. But, as we will see below, this effort ran into trouble for at least two reasons:

first, it did not take into account broader moral and political debates about the constitution of human groups in the social order; and second, it did not take into account broader debates in biology about whether meaningful human groups exist in nature.

GROUPS CAN BE DEFINED FOR THE PURPOSES OF CONSENT

The second assumption that underlies the Model Ethical Protocol's group-consent provision is that groups could be defined for the purposes of consent. But defining the contours of a group that could legitimately give consent quickly became a problem for the Protocol. One drafter, for example, raised the following problematic set of questions:

> Is the population this village? Is it the "Plains Apache"? Is it all the Apache? Is it all Na Dene speakers? Is it Native Americans? What's the population? Second, is it a population for whom consent can appropriately be sought? Is it a population where you can meaningfully ask either the group as a whole or some representative body for permission? Is it Irish Americans or is it a Native American tribe? Is there a structure there that one can communicate with, or is it so small that you can actually get the entire population in a group meeting and let a consensus emerge or not? Then at the third level . . . if it is too big to have everyone in a room talking, but there does seem to be a structure within the population that provides for the possibility of getting consent, who do you get the consent from? (Interview Z)

To answer these questions, the Protocol settled on two definitions of groups, each sanctioned by an expert community.

The first definition of group drew upon the expertise of Diversity Project researchers. As the Protocol states,

> If the researchers intend to sample only a particular part of such a unit [town, village, or religious unit], *the group* from whom consent must be sought is defined *by the researchers'* sampling criteria. For example, if researchers wanted to sample all members of a village who spoke a particular language, that portion of the village would be the relevant community [italics added]. (North American Regional Committee 1995, 9)

Here researchers are placed in the position of defining groups using their sampling criteria as a guide.

The Model Ethical Protocol also draws upon an expert-defined unified notion of culture to define groups. Groups, it states, are "distinct cultural entities" (ibid., 9). It draws upon a notion of tradition to define these cultural entities. In the Protocol, "tradition" distinguishes an imagined group of insiders to the sampled population from "the outsiders." It is an object that scientists must respect, and not violate. This notion of the sampled group as a "traditional culture" that must be respected is perhaps most vividly illustrated in the Protocol's section, "Before Contacting the Population," which presents the following scenario:

> In many societies around the world, hair is secretly collected from intended victims to harm them through witchcraft. Consequently, people collect their own loose hair, fingernail parings, and other body products and bury them to avoid this danger. Researchers who asked such a population for hair might be seen as intending to perform witchcraft. Blood is often intended as a sacrifice, sometimes through special rituals. (ibid., 5)

This hypothetical problem is followed by the following potential solution: "Donation of blood in such cultures is a serious matter that would require discussion and perhaps a neutralizing ritual" (ibid., 5). This prescription presumes the existence of a culture that might be "neutralized," roped off, and protected from the realm of biology.

The Protocol assumes that knowledge of this brand of culture resides in experts and can be drawn upon to resolve some of the complex identity questions group consent raises. For example, the Protocol explains, the Maori of New Zealand are organized into groups called iwi and present the problem of determining whether or not a pan-Maori organization exists that all iwis would recognize as a "culturally appropriate authority" that could give consent. For them, the Protocol suggests, "researchers would need to have a very sophisticated knowledge of Maori culture and politics before proceeding" (ibid., 10). It is presumed that expertise about Maori culture and politics will resolve the complex questions about who speaks for the Maori.

The Protocol does not assume that the culture of groups will be clear to Diversity Project researchers. To the contrary, it assumes that biological researchers will have to be taught to understand and see these cultures by what it deems to be cultural experts. Thus, it directs researchers to consult with "anthropologists, or others knowledgeable about the population" before beginning their research plans. It also notes that the North American Regional Committee would "be happy to recommend experts on particular populations" (ibid., 11).

GROUPS DEFINED BY WHOM AND FOR WHAT PURPOSES?

The assumption that groups can be defined by experts elided the broader set of concerns expressed by RAFI and members of indigenous rights organizations in the summer of 1993. As described above, these groups worried that expert definitions of indigenous groups would construct their identity in ways that would threaten their security and autonomy.

Diversity Project organizers' role in constructing the social identity of groups while defining the objects of research, however, proved difficult for organizers to see. Drafters believed that they were only attempting to sample already existing groups in an ethically responsible way. They assumed further that appropriate experts could identify these already existing groups.

That this might not be the case became clear as members of tribes in the United States were called upon by government agencies to review the Model Ethical Protocol. Now the question became who could decide what constituted informed consent. This question was more basic than mere compliance with ethical codes. It connected to fundamental issues about tribal identity and sovereignty rights, for in order to get consent from a group, one first had to define the group and determine who had the authority to speak for it. Who would have this power became the critical question.

As one Native attorney explained, it looked like the Model Ethical Protocol ultimately placed this power in the hands of the North American Regional Committee of the Diversity Project. This Committee, at least as described by the Protocol, would decide if a sampling effort had followed the ethical guidelines (like group consent), and thus could be admitted as a part of the Project (North American Regional Committee 1997, 1435).

This allocation of power to the NAmC raised an obvious problem for tribes in the United States: namely, it threatened their sovereignty rights. As a Native attorney in health policy explained:

> It is a sovereign right of peoples to define their own members. This is a right that tribes in the U.S. have had from time immemorial. The United States protects this right for federally recognized tribes. The right has been attacked, and upheld, in the United States Supreme Court. Other times when others have attempted to impose a definition on group (tribal) membership, including the federal government, the effort has been motivated by political or economic expediency. Tribal membership is one of many sovereign rights that outsiders have attempted to circumvent or deny native people, and every time their attempts have proven divisive and very harmful if actually put into practice. Other attempts will be similarly harmful. Only native people are able to make these decisions, not scientists, lawyers, or ethicists. (Interview A1)

Court cases such as *Santa Clara Pueblo v. Martinez* upheld the rights of tribes to determine their own membership (Wilkins 1997, 214–15; Cohen 1982, 246–48). By placing the power to determine group membership and voice in the hands of experts such as anthropologists, lawyers, and ultimately, ethicists, the Diversity Project threatened to encroach upon these rights.

Decisions about who defines groups and how matter deeply in ways that may not be immediately apparent. Group membership and definition help arbitrate who will and who will not qualify for federal funding and aid. And this is to say nothing of the cultural values and identities at stake. Members of tribes asked to review the Diversity Project believed that the power to define group identity—with all its possible consequences—should not be left in the hands of a few scientists who might want to study human genetic diversity, or the anthropologists, lawyers, and ethicists who would facilitate their project. Although they did believe group consent had been proposed in the right spirit—to secure the rights of groups—it faltered as it failed to address the complex questions about power that underlie efforts to define and protect the rights of groups.[193]

In addition, some Diversity Project organizers themselves worried that the Model Ethical Protocol's group consent provision would reintroduce colonial understandings of indigenous groups. Why, they argued, had indigenous populations been singled out as groups from whom consent would be required? As one organizer explained: "People collect data on Blacks all the time, every place, all over the country, and nobody has held a national forum of Blacks . . . [but now that we are talking about indigenous populations] . . . now suddenly the issue gets raised" (Interview A2). So, why, this scientist asked, do we treat indigenous groups differently?[194] To do so, he continued, constituted a form of paternalism: it treated indigenous groups as if they were childlike and in need of protection. Far from being ethically innovative, he argued that group consent threatened to reinscribe old racist notions of indigenous groups as primitive and isolated and in need of protection. None of the groups the Project hoped to sample—such as tribes in the United States—fit this stereotypical model: "They don't live in a little village. They are not on a discrete reservation. They are all over the place. They have deep factions themselves, like everybody. And so you don't get [them] together, like they were some sort of savages" (Interview A2). In short, some argued that group consent represented a kind of paternalism that was the very opposite of what indigenous groups had demanded.

BEYOND THE LIMITS OF WESTERN BIOMEDICAL ETHICS?

As this account demonstrates, group consent raised fundamental political questions about how and whether to define groups for the purposes of granting them rights. Diversity Project organizers, of course, were not alone in their struggles to address these questions. Indeed, questions raised by group consent strongly resonated with those raised by another attempt to move beyond Western liberal democracy's focus on the individual: communitarianism. Communitarianism challenges the prevailing political philosophy in the United States, liberalism. Liberals such as Immanuel Kant, John Locke, and John Rawls argue that governments should protect the rights of individuals and assure their ability to choose their own ends. Communitarians, like the political philosopher Michael Sandel, on the other hand, reject the liberal notion that individuals are autonomous and independent, arguing instead that all individuals are at least partly constituted through bonds with others.[195] Indeed, these critics of liberalism hold that the individual's capacity for self-determination depends upon his or her ability to participate in a political community. This community provides a shared discourse through which citizens engage in the rational and peaceful debate required to secure liberty and democracy.[196] As Sandel explains in *Democracy's Discontents*:

> On the liberal view, liberty is defined in opposition to democracy, as a constraint of self-government. I am free insofar as I am a bearer of rights that guarantee my immunity from certain major decisions. On the republican view, liberty is understood as a consequence of self-government. I am free insofar as I am a member of a political community that controls its own fate, and a participant in the decisions that govern its affairs. (Sandel 1996, 26)

NAmC's group-consent provision resonated with this communitarian principle that communities that have the power to control their own fate are vital to democratic self-realization.

Attacks on the NAmC group-consent provision, not surprisingly, are reminiscent of criticisms of communitarian political theory. By assuming the existence of a relatively homogeneous community with a shared discourse expressing shared values, critics argue communitarians fail to adequately recognize differences within political communities (Smith 1998, 124). Sandel, in particular, has attempted to address these concerns by acknowledging that an individual will often have multiple and overlapping memberships, and by calling for a society in which multiple interpretations of the "common good" would be maintained (Sandel 1996). Even with these revisions, however, post-structural political theorists such as Ernsto Laclau and Chantal Mouffe find lingering in communitarianism a more serious problem: the theory provides almost no account of power relations. Although recognizing differences, Laclau and Mouffe argue that Sandel and others assume that enough of a common way of life will emerge from rational discourse and debate to allow for the peaceful resolution of these differences. This assumption, they continue, prohibits recognition of a fundamental fact of contemporary social groups: they are complex and hybrid; they do not merely exist, but rather are defined and constantly re-defined through political contestation; at stake in these contestations are fundamental issues of power, including access to material resources and the ability to self-determine. The assumption of cultural homogeneity elides these important issues of power (Smith 1998, 125).

An analogous problem troubled the Diversity Project. Organizers assumed that critics would become supporters if they participated in a rational and informed dialogue about the Project's intentions and goals. They had difficulty recognizing how specific histories of colonialism and racism, and legacies of Western domination, might interfere with the possibility of such a dialogue. In particular, drafters of the Model Ethical Protocol did not recognize that for historically oppressed groups the very effort to gain group consent would raise questions about who defines groups and for what ends. This lack of attention to historical conditions and fundamental questions of power called into question the democratic potential of group consent.

Conclusion

As this chapter has demonstrated, in the face of a second wave of criticism by nonscientists, the scientists who proposed the Diversity Project looked to the norms and practices of Western biomedical ethics to stabilize their Project. In particular, they sought to create informed subjects who could give their consent to participate in the initiative. In this way, they hoped to draw upon already established ethical practices to accomodate their novel scientific initiative. However, on the topic of human diversity, perhaps the most fundamental and contested in history, those deemed ethical experts proved no more able to construct a stable moral order than those deemed scientific and cultural experts proved able to make definitive claims about natural order. These expert-based approaches faltered as questions about moral order proved inseparable from questions about the order of nature and the constitution of power.

No other episode in the history of the Diversity Project debates would illustrate this better than group consent. Developed in Western contexts where the subject of rights and the locus of autonomy is assumed to be the individual, the principle of informed consent—despite the efforts of the NAmC to adapt it to groups— simply could not provide the tools needed to think through and address the intertwined political and scientific issues at stake in determining the status and meaning of groups in nature and society. The group consent provision of the Model Ethical Protocol merely assumed that Diversity Project researchers could provide answers to the complex questions about the biological definition of groups, and that "anthropologists and or others knowledgeable about the population" could provide answers to an attendant set of questions about the definition of groups in culture and society (North American Regional Committee 1995, 9, 11). By turning to what its authors deemed scientific experts on the one hand and cultural experts on the other, this proposed expansion of informed-consent rights bypassed vital questions about the status of groups as entities in nature and society, as well as historically embedded and contentious questions about the role that racial and ethnic categories could play in ordering studies of human biological diversity. As the group consent case demonstrates, although sophisticated notions of human groups have evolved in other contexts, these notions have not as yet been transported to the field of Western biomedical ethics.

The fear for many was that in the absence of more sophisticated understandings of human groups, the ethical tool of group consent might unreflectively reinscribe old colonialist, paternalist, and reductionist definitions of groups. In particular, some worried that the NAmC's group-consent provision would reintroduce racist and colonialist notions of indigenous groups as isolated and primitive, and that these understandings would be both scientifically and politically destabilizing. In my concluding chapter, I will return to the problems of group consent, and what they reveal about the limits of Western biomedical ethics. Before exploring these broader implications, however, I turn to the second major response to critics proposed by drafters of the Model Ethical Protocol: the participation of diverse populations, including "major ethnic groups," as both knowing subjects and objects of study.

Chapter 6
Discourses of Participation

From the Diversity Project's inception in 1991, proponents argued that their sampling initiative would end the "Eurocentric bias" of the Human Genome Project (Bowcock and Cavalli-Sforza 1991, Human Genome Organization 1993). To date, they observed, most studies of human genetic diversity had "been made on Caucasoid samples for obvious reasons of expediency" (Bowcock and Cavalli-Sforza 1991, 491). The Diversity Project would correct this bias by sampling indigenous populations around the world. Some geneticists and biological anthropologists who identified themselves as African American, however, did not agree.[197] They contended that by including the genomes of indigenous populations, but not those of African Americans, the Diversity Project would further support structures that propagated the exclusion of relevant minorities from genomic research (Jackson 1998).[198]

Project organizers responded to these criticisms by invoking rhetorics of participatory democracy. The initiative, they argued, would be participatory in the fullest sense. First, it would include "all human populations," including "major ethnic groups" such as "African Americans" (Weiss 1993; North American Regional Committee 1994c, 6). Second, it would include all populations not only as objects to be studied, but also as "partners" in the research process. As the Model Ethical Protocol's "Partnership with Participating Populations" section explained, these partners would both design and conduct Diversity Project research (North American Regional Committee 1995, 28–31).[199] In other words, populations would be more than mere objects of research, and more, even, than ordinary human subjects; they would stand alongside the scientists as active producers of knowledge. In this way, organizers hoped to build a participatory and democratic initiative—one that would, in the words of one prominent organizer, follow in the political legacy of American affirmative action (Weiss 1993).

Members of the North American Regional Committee expected that their critics would welcome these proposals. Many, however, responded with new worries. In particular, some questioned whether these new participatory measures would address community needs, or merely respond to pressures potential funders placed upon Project organizers to demonstrate that they had adequately represented and included groups in their initiative.[200] Many suspected the latter. Rather than engage with their concerns, some critics worried that Project organizers' efforts to generate participation would act to legitimate it through "exhibitions of representation" (Amit-Talai 1996).

These worries about the different effects of participatory measures emanated from a long history of political struggles over the meaning of participation in the United States. While some, most prominently leaders of the Civil Rights movement, pushed for the inclusion and participation of minorities in American institutions, others, including some prominent African American leaders, and many members of American Indian tribes, viewed inclusionary efforts as a form of co-optation. By treating participation in an ahistorical, apolitical, and model-like manner, some worried that MEP's partnership provision failed to address these ambiguities and tensions in discourses of participation as they have developed in the United States, and did not acknowledge the role the Project might play in reshaping the meaning and practices of inclusion in the American polity.

Further, as in the case of group consent, rather than including already-formed groups in an ethical and democratic manner, critics worried that efforts to generate partnerships would in effect produce groups in society in a manner that would reproduce structures of objectification and co-optation. Specifically, those asked to review the Model Ethical Protocol (MEP) for tribes in the United States noted that the partnerships provision drew upon a model of knowledge and power that located knowledge with the scientists. On MEP's account, in order for members of populations to participate in the design of research questions, scientists would first need to disseminate knowledge to them so they would know "what is possible" (North American Regional Committee 1997, 1470). Far from empowering the recipients of knowledge, some raised concerns that this one-way flow of knowledge (from researchers to populations) would only serve to render populations the object of power (by, for example, creating the threat of their co-optation).

Drawing upon a framework that cast the Project as merely attempting to represent already formed voices and groups in society in a democratic manner, the Project's proposed participatory measures fell short of providing the tools needed to address critics' concerns. Further, by relying on a conception of science that opposed it to society, questions never emerged about how the Project's claims about socially meaningful groups, like African Americans, might tie to particular definitions of groups in nature. Proponents of the participatory approach merely stated that "all human populations" would be included; they made no mention of the earlier Project debates about how populations could be defined, and the role, if any, socially meaningful categories of race could play. This caused concern among some organizers of the Project in Europe who argued that major ethnic groups in the United States could reveal little about the evolution and diversity of the human species, and thus would make inappropriate objects of study given the questions about natural order that originally motivated the Project's proposal.

This chapter brings into view these worries about the role organizers' procedural choices might play in simultaneously producing the Project's objects of study, and the subjects who could speak for these objects. It demonstrates that Project organizers' decisions about how to define groups in society who could participate in human genetic diversity research could not be made independently from decisions about how to define groups in nature that could meaningfully represent human genetic diversity. Rather, each set of decisions was implicated in the other. Organizers' efforts to stabilize their Project by turning to what they viewed as the ideals of democracy, such as participation, would prove unsuccessful as the problem of ordering and representing human diversity for the purposes of democratic practice and scientific research proved inseparable.[201] Solutions to problems of order, legitimacy, and authority raised by efforts to study human genetic diversity require asking more fundamentally about the mutual constitution of scientific research and democratic political practice.

Discourses of Participation

Despite their many differences, organizers of the Diversity Project did agree on one thing: the Project would not move forward if people refused to participate. However, understandings of who should participate, and how, changed over time.

At first, an instrumental goal guided the researchers' decisions: they wished to gain access to the largest number of "isolated human populations"—populations deemed to have the most "informative genetic records" (Cavalli-Sforza et al., 1991, 490). To reach these populations, organizers did not in the first instance seek their direct participation, but rather the participation of "anthropologists, medical researchers, and local scientists who already have access to the more isolated groups." At this point in the evolution of the Project, proposers did not anticipate that they would directly interact with populations. Rather, most expected that those who already had the necessary "working relationships" would facilitate DNA collection. Thus, not surprisingly, organizers focused most of their efforts on enrolling these professionals, not the populations themselves.[202] Any thought devoted to figuring out how to encourage the participation of populations drew upon past examples of providing basic health care. As *Science* reported: "Already, says [Mary-Claire] King, the group is thinking of what it can offer to the populations in return, such as medical supplies" (Roberts 1991b, 1617).

However, following charges by physical anthropologists in 1992 and early 1993 that the Project might re-ignite biological racism, organizers' ideas about who should be included, and how, began to change. In particular, some began to recognize the necessity of addressing broader issues of racism and the politics of inclusion. They acknowledged that those who originally proposed the Project were all white Europeans and Americans, and if they did not want the Diversity Project to be viewed as a Project organized by "colonialists," the composition of its leadership would have to change (Interview A4).[203]

The new goal of generating nonwhite and non-Western support and leadership for the Project prompted supporters to organize an international planning meeting for the fall of 1993. As one member of the Diversity Project's North American Regional Committee explained:

> [W]e had decided at the Washington [Bethesda, Ethical, Legal and Social Impliations Program (ELSI)] meeting that there should be a world meeting and that people from developing countries should be invited and get a say as to what should go on if there was going to be such a project. So that was the purpose of the Sardinia meeting—to get people from Japan, from Africa, from India, places where some sampling was likely to go on . . . to have those people involved in establishing a structure that might operate the Project for the world (Interview B).

This international meeting held in Sardinia, Italy, took place seven months later.[204] The new understanding of participation arose from the effort to link the Project to a different discourse of participation—one shaped by policies of inclusion in the United States, most notably, affirmative action. Although it was understood as a response to what organizers deemed political pressures, this new notion of participation also carried along with it a different understanding of what they saw as the Project's scientific goals. New objectives and goals of research emerged along with the new definitions of research subjects.

CHAPTER 6

New Models of Participation

Efforts to overcome racism in the United States have largely been defined by the exclusion of African Americans from societal benefits: in particular, education, jobs, and the right to vote. Affirmative Action, a policy issued by presidential executive order in 1965 to enforce the Civil Rights Act of 1964, became the leading means to overcome discrimination, and to achieve the full inclusion of historically oppressed groups in all parts of society. During the late 1980s and early 1990s, scientific research—in particular, biomedical research—emerged as a new site for this old battle for inclusion (Epstein forthcoming).[205] Critics argued that racist forces shaped science just as much as any other dominant social institution. By the mid-1990s, some African Americans in public health and biomedical research began to demand that these inequities be addressed (Mittman 1997, Freeman 1998).

To understand how these broader debates about the inclusion of racial and ethnic minority groups in health care and biomedical research shaped the Diversity Project, it is important to look closely at two other projects that sought to include African Americans in biological and biomedical research: the Genomic Research in African-American Pedigrees project (G-RAP), and the African Burial Ground Project (ABGP).

GENOMIC RESEARCH IN AFRICAN AMERICAN PEDIGREES

In 1990 a group of geneticists and biostatisticians at Howard University, a historically black college, proposed a project to make genomic research more directly useful to African Americans. In particular, they sought to create a reference set of African-American families to complement the existing set of Caucasian families used by the Human Genome Project, and set up by the Centre d'Etude du Polymorphisme Humaine (CEPH) in Paris (Sankaran 1995).[206] Supporters of G-RAP argued that without a specific accounting of genetic variation by race, the Human Genome Project would lead to an exacerbation of racism.

The Howard team received a two-year planning grant in 1991. At the end of this time period, Georgia Dunston, a human geneticist involved in the project from the beginning, proposed to expand it to a large-scale effort to develop a linkage map of the genomes in African Americans. Playing on "rap," an African American colloquialism for "talk," the project became known as G-RAP (i.e., genomics "talk" or "let's talk genomics"). As Dunston explained in a *New Scientist* article in 1995: "We need to expand the reference genome to make it relevant to all populations. . . . We know enough about [human] evolution and genome diversity to conclude that Africans are the most genetically diverse people" (Sankaran 1995). This greater diversity, she argued, meant that many of the variations that correlate to diseases in African Americans would go undetected if genome research were conducted on only Caucasian DNA. The G-RAP project sought to identify and locate these variations.

Other African American researchers and health-care professionals supported Dunston's argument that the Human Genome Project needed to be complemented by a project that characterized the genomes of those of African descent.[207] Unless efforts were made to include these genomes, the biological anthropologist Fatimah Jackson asked:

> How will HGP information benefit Americans of African descent and other citizens of color who have shared in the expense and sacrifice for this highly touted project? How will their inevitable variation from the Eurocentric molecular baseline generated by the HGP be addressed? (Jackson 1998, 163)

Ilana Suez Mittman, a professor of genetic counseling at Howard University, argued that in order to take into account this variation, and to counter the view that genetic research was simply "another form of oppression," "ethnically diverse groups must be properly included and actively participate in the field of human genetics" (Mittman 1997).

In the context of these broader events and concerns, it becomes clear why some considered the Diversity Project as another warning sign that African Americans were being left out of the genetic revolution. The first proposers of the Project neither planned to sample African Americans, nor to include them in the planning of the Project. Indeed, Dunston, the only African American to become involved, heard about the sampling initiative only accidentally.[208]

Early Diversity Project organizers did not merely fail to include African Americans in the Penn State list of "recommended populations"; they viewed these populations as the antithesis of the kind of population they hoped to sample. According to these proponents, the genomes of "more recent, urban" populations contained less information of historic interest, and thus were not of direct interest for the purposes of the proposed sampling initiative (Cavalli-Sforza et al., 1991, 490). However, in the wake of criticisms issued by African-American scientists and health-care providers—in particular, those who had not been invited to the second planning workshop—organizers' claims about the scientific merit of studying African Americans began to change. Now, some leaders of the Diversity Project argued that these populations were the very ones the Project wanted to serve. As Kenneth M. Weiss, soon to be chair of the North American

Regional Committee of the Diversity Project, explained in a letter to the U.S. Congress in April 1993:[209]

> This country has among its citizens numerous people from all but the smallest and most isolated human populations. Our big cities like New York and Los Angeles, and our states, like Alaska, Arizona, and Hawaii, are inhabited by very diverse populations, and these include large numbers of people with admixed ancestry among different populations. One need think only of Mexican-Americans, with European-Amerindian ancestry, or African Americans, to remember that there are *tens of millions* of these people in our population. In aggregate, they may soon comprise the majority of our citizens. For various diseases, these people are subject to risks that differ, probably at least in part for genetic reasons, from the risk in the better-studied part of our nation [emphasis in original]. (Weiss 1993, 44)[210]

On these grounds, Weiss advocated a "worldwide sampling of human genetic diversity, including rare and endangered populations *as well as the major ethnic groups* [italics added]." This, he argued, would involve sampling "populations that comprise our major ethnic groups (proper representation of Africans, Amerindians, Pacific and Asian populations, Indians, etc.)." He concluded that such a sample would provide "biomedically critical data" (ibid., 44).[211]

To justify this vision of sampling, Weiss invoked U.S. efforts to include racial and ethnic groups as full participants in society, in particular the government's affirmative action policies. As he explained:

> The National Institutes *mandates affirmative action* in all of its research grants, to ensure that all of our nation's people are served by the research and clinical establishment. How is it that we can allocate vast sums of money to studying human genes in a way that specifically *excludes* consideration of relevant diversity? [italics added]. (Weiss 1993, 44)[212]

This exclusion, Weiss continued, was not only contrary to NIH policy but also a determinant of the future of biomedical science. He hoped that the Diversity Project would ensure that all American minorities were included in this future by including their genetic variants in any database of human genetic variation. Indeed, he believed that G-RAP would be part of the Diversity Project (Interview 1996).

This hearing document represented one of the first signs that sampling African Americans and Mexican Americans—racial and ethnic groups officially defined and recognized by the United States Census Bureau—would be a Diversity Project goal. Inviting Georgia Dunston to be on the North American Regional Committee in the fall of 1993 was another.

Further changes to the Project's notions of who would participate in the Project—and how—resulted from exchanges with the director of a second effort to employ human genetics to serve the interests of African Americans: the African Burial Ground Project.

THE AFRICAN BURIAL GROUND PROJECT

In 1991, in the course of breaking ground on a new building in New York City, the General Services Administration (GSA) "rediscovered" the African Burial Ground, the main cemetery for Africans in colonial New York City. As required by the National Historic Preservation Act, an excavation of the 427 remains began. The main goal was to remove the skeletons that impeded construction of the building. According to published accounts, the project provoked the anger of local African American communities, who charged that the GSA had ignored their demands to play a principal role in the disposition of the remains. In response to the protests, the city's mayor and the U.S. Congress called the project to a halt. Later that year a group of researchers at Howard University led by the biological anthropologist Michael Blakey were awarded the contract for analysis of the site. Their effort became known as the African Burial Ground Project (Blakey 1998).

Blakey, as scientific director, instituted a process of "engagement" with the "community" (Blakey 1998, 400). As part of this process, those working on the ABGP distributed copies of the 130-page research design to all interested persons and held hearings in downtown Manhattan and Harlem to receive feedback. The goal was to make sure that the project reflected the interests of the "descendant or culturally affiliated community" by asking questions that were of interest to the community and also "appropriate for a scientific program."[213] African Americans were also included on the epistemological grounds that they would "bring a perspective that might help rectify the distorting effects of Eurocentric denial of the scope and conditions of African participation in the building of the nation." According to Blakey, through this process, members of the "African-American public" came to "own" the project.[214] He concluded, echoing the famous finale of Lincoln's Gettysburg address, "Ours is a study of the people, by the people, and for the people principally concerned" (ibid., 400).

In November 1993, when the Wenner-Gren Foundation meeting on the Diversity Project convened in Mt. Kisco, New York, the ABGP had just begun.[215] At the Mt. Kisco meeting, Blakey shared his experiences fostering "public engagement" in the new project. As one of the Diversity Project's organizers put it in an interview, Blakey's talk inspired the second change in Diversity Project organizers' ideas about who would participate in the Project and how (Interview P).[216] Instead of viewing populations as mere human subjects who had the power to give or not give their consent to be sampled, after Mt. Kisco members of the North American Regional Committee began to envision populations as partners in the research process who would not only provide DNA samples, but also formulate research questions, design sampling strategies, and collect samples.

The new approach to defining research questions and the role of research subjects introduced by Blakey found formal articulation in the "Partnerships with Participating Population" section of the Diversity Project's Model Ethical Protocol.[217] As the Protocol explains:

> The key word is "with." We believe that, ideally, researchers involved in collecting samples for the HGD Project should be closely connected with the populations that provide those samples, connected not as "scientists" and "subject," but as partners. They might even be the same people, as communities may undertake to sample themselves in order to participate in the Project. (North American Regional Committee 1997, 1468)

Just as Michael Blakey, an African American, had been enlisted to study the remains of African Americans, so Diversity Project organizers argued that members of different racial and ethnic groups should study their own genes: "Indians work with India. We have no American Natives . . . and we don't have a Black person working on Africa, but we have . . . Asians working on Asia, for instance—not only Indians, but also Chinese and so on" (Interview E). Another organizer defended the sincerity of this proposal:

> [We] would not just be sneering at these people. I don't think we are pretending—I think we would love to have a Kenyan in our lab, or a Yanomama in our lab if we could get a qualified person, or a scientist from Brazil in our lab. Or help set up a lab in Brazil. (Interview A2)

And indeed, organizers of the Diversity Project did make efforts to both enroll and train members of populations they identified as of interest for sampling.[218]

INSTRUMENTALISM

Although explicitly modeled after Blakey's notion of "descendant or culturally affiliated communities" in the ABGP, the Protocol's partnership provision differed from Blakey's in at least two ways. First, in the case of the ABGP, those participating in the design of the research were not themselves research subjects. Blakey was not asking for blood or tissue samples from members of the "descendant or culturally-affiliated communities." In the Protocol, however, the group representatives were at the same time advisers and research subjects. This made the political and ethical issues at stake in the Diversity Project significantly different from those in the ABGP.

Second, Blakey's collaboration with concerned ethnic groups derived from the belief that it was ethically and politically the right thing to do. The Model Ethical Protocol, on the other hand, proposed partnerships both because organizers believed that ethically this was the right thing to do and because they believed that partnerships would help generate support for the Project.[219] For example, the Protocol explicitly noted that partnerships could overcome the problem of real or perceived exploitation:

> Sampling for the HGDP will involve numerous tasks. . . . We recommend that, whenever possible, the researchers use local people in performing these tasks. These activities will allow the local population to be more engaged in the research. Their involvement in the collection process, like their involvement in the planning, should also prevent the reality or perception that the researchers are exploiting or abusing the population. (North American Regional Committee 1997, 1471)

This passage gains significance if we recall that at this time organizers regarded the perception of exploitation as a major cause of opposition to their Project.

More specifically, the Protocol explained that partnerships would enable researchers to give something back to populations. Indeed, the need to create a reciprocal exchange became one of the two reasons given for revising and expanding the normal model of interacting with research subjects so that it included partnerships with sampled populations. As the proposed Model Ethical Protocol put it:

> The community may have specific questions they find interesting. . . . Researchers need to look for these questions for several reasons. *First*, the community's questions are entitled to respect because the community is interested in them. It is allowing the researchers to take information important to its identity and fairness dictates the use of that information, where possible, to answer the community's questions. (North American Regional Committee 1997, 1470)

In such passages, the Protocol explicitly acknowledged that researchers want something from populations, and to create a fair exchange (and avoid feelings of exploitation), partnerships would be useful.

Such passages also recognized that partnerships could help scientists design research questions: "A community necessarily knows things about its history and health that no outsider, no matter how expert, can know. That knowledge may lead to questions of substantial scientific issues, questions the researcher would never know to ask" (ibid.). Yet, statements like the following made it clear who would play the primary role, namely, the research community:

> Of course, the sampled population is unlikely to present researchers with fully worked out, scientifically testable hypotheses. The process will be a complex one, with the community explaining its interests, the researchers explaining *what is possible*, and then both sides starting again on the basis of that discussion [italics added]. (ibid.)

Researchers would define what was possible. There is no suggestion that members of populations might also be in a position to tell researchers what was possible.

In addition to involvement in the planning and collection process, the Protocol argued that populations should be involved in the results of the research. Again, many of the reasons given were instrumental: "Too often, participants in research projects have no ongoing involvement after donating their samples and time. This might generate resentment; it will inevitably waste an opportunity to get the public more involved in, and knowledgeable about, science" (ibid., 1472). In this passage, as well as many others, the authors of the Model Ethical Protocol focus on why partnerships are good for researchers and for science: they could help reduce resentment and enroll the public in learning about science.[220]

THE IMPORTANCE OF HISTORY

This instrumental approach to partnerships would soon prove problematic. We can begin to understand why by noting the ahistorical manner in which the Partnership provision of the Model Ethical Protocol was presented. As described above, the notion of partnership developed in the Model Protocol arose within a specific history of race relations in the United States. Indeed, many recent approaches to democratic participation in the United States emerged out of the political struggles of African Americans against the enduring legacy of centuries of slavery. In the Protocol, however, there is no mention of this history, or even of Blakey and the ABGP. This context is erased and a depoliticized version of partnership is presented as if it can act like a neutral procedural tool.

This decontextualization is one of many indications that organizers of the Diversity Project faced difficulties recognizing the importance of the histories and political struggles from which their procedural choices arose. Instead, they attempted to extract norms and practices from other contexts and apply them already formed to their Project. In so doing, they failed to engage with the ambiguities and tensions of participation in American institutions in general, and the Diversity Project in particular.

The problems with this approach first became apparent as organizers of the Project began to use their notion of partnership to address the concerns of members of tribes in the United States asked to participate in discussions about the Diversity Project. What they would perhaps all too slowly learn in the process is that the history of African Americans and the history of American Indians differed in ways that mattered. As Frank Dukepoo, a Hopi geneticist who had been involved in the Diversity Project debates from a very early point, later explained:[221]

> Although African Americans and Indians are members of the same genera and species and share the same number of chromosomes, as groups we differ markedly in origin, culture, customs, language, experience with the United States government, and worldview, all of which shape our thinking and values. For example, American Indians have a history of more than 500 years of colonization that Blacks do not. In addition, Indians have land bases called reservations, rancherias or reserves and Blacks have no comparable base. Indians also have a special legal relationship with the Federal Government but African Americans have no such relationship. Furthermore, Indians have entered into treaty-making while Blacks have no such history (Dukepoo 1998).

Within this particular history—one marked by colonization, broken promises and forced assimilation—participation in dominant society in the United States carried a different meaning for many American Indians than it did for most African Americans.[222] Interaction and participation had led to further loss of land, identity, and dignity. Thus, most did not seek participation, but rather separation and self-determination.[223]

Science was not exempt from this history. As Dukepoo explained, "formal exploitation" might have ended with the final defeat of Geronimo in 1886, but colonization continued in a more "benign" form.[224] Pursuit of knowledge, including education and scientific research, played central roles in this second phase of colonization:

> During the 20-year period between 1890–1910, Indians reached the nadir of their existence as they were confined to reservations. On the verge of decimation, the population reached an all time low of just under 250,000. Indian children were forcibly removed from homes and made to attend private and government boarding schools where they were forbidden to speak their native languages and practice their cultures. In a more desperate attempt at assimilation, the children's hair was cut to make them look like whites. The prevailing philosophy is

> depicted by a boarding school administrator at the time: "Kill the Indian but save the child." Meanwhile, indignity abounded as Indians served as research subjects in numerous social, anthropological, political, educational and medical studies and experiments. Currently, a large number of Indians are being studied. (Dukepoo 1997, 4)

Indeed, he continued, so much research had been done that "[a]lthough a few Indians would willingly get involved in genetic research, many express the feeling that they simply are tired of serving as research subjects" (ibid., 1998, 3).

Many viewed the study of human genetic diversity as part of this long legacy of scientific research that had led to further colonization and assimilation.[225] For some, Dukepoo argued, the only difference between genetic research and previous studies was that genetic research had "turned up the burner" on old concerns. Dukepoo even went so far as to dub genetic research the "Second Coming of the Whiteman." As he explained:

> In the first coming, whites arrived in ships; now they are arriving with biotechnology. While some Indians quaintly refer to the new era as the "Biotechnology Revolution," others deem it "Biocolonialism." For the latter, advanced technology applied during colonization enabled the Whiteman to "develop," extract and exploit the natural resources of native land. They contend that whites were infected with and still suffer from "land," "gold," "uranium" and other fevers. Today their concern is mounting as "green fever" infects Western scientists who seek plants for medicinal products along with the indigenous knowledge of their application and use. . . . They also assert that human geneticists, medical professionals, anthropologists and evolutionists have severe "Indigenous Fever" as they await the outcome of National Research Council's deliberation of the Human Genome Diversity Project. In their minds, Western scientists literally are after Indian blood. (Dukepoo 1997, 12)

In the context of this economic and political subjugation, not surprisingly, offers to participate in the design and conduct of genetic research were not met with enthusiasm.

Differences in the meaning of participation existed not only between major ethnic groups, but also within them. Although some African American scientists, such as Dunston and Blakey, pushed for the inclusion of African Americans in genetic research, others proved more wary. Referring to herself as the Malcolm X of genetic research, one African American biological anthropologist questioned the Diversity Project's connections to the Human Genome Project, an effort she considered a manifestation of the interests of the dominant society (Interview 1999). Her views reflected a strong tradition of separatism within African American communities in the United States (Malcolm X, 1964). Caution about participating in biomedical research also stemmed from the legacy of the Tuskegee Syphilis Experiment (Jones 1981). As we will see below, these different historical positions proved central to the Diversity Project debates about participation and partnership.

IN WHOSE INTEREST?

Challenges to the belief that participation in the Diversity Project (as both project designers and research subjects) would benefit studied populations arose at the first Diversity Project meeting supported by the MacArthur Foundation held in San Francisco in spring of 1994.[226] At this one-day event, entitled The Human Genome Diversity and the Interests of Native Americans, representatives of the NAmC introduced Blakey's model of partnership. It quickly became evident that this model grew out of a very particular history of race and racism that was not shared by all.

First, whereas Diversity Project organizers could plausibly argue that African Americans wanted to discover their history and roots (by citing the ABGP, for example), one could not make a similar claim about members of tribes in the United States or First Nations in Canada. No analogous demand for origins research existed. To the contrary, those present at this first MacArthur meeting expressed an explicit lack of interest in this kind of research. As one participant in the dialogue between organizers of the Diversity Project and those asked to represent Native Americans explained:

> We are not interested in you proving the land bridge because we have stories already that tell us that it is not true. We question the validity of certain scientific procedures like carbon dating and can prove to you on your own terms that is not accurate. We don't need linguistic studies because we know who talks like us, and so if it is all those things, then we don't want it. (Interview F)

The Out-of-Africa theory, the scientific theory that all humans originated in Africa, did potentially form a bridge to one tenet of the indigenous rights movement—that all indigenous people were connected (Interview M; Akwasasne Notes 1978). But beyond this, genetic research into human origins did not interest those asked to represent the interests of Native Americans. Questions about the relationships between the indigenous peoples of the Americas, they argued, had been answered long ago.[227]

Some even found organizers' attempts to forge links to their stories about origins insulting. At a subsequent meeting also sponsored by the MacArthur Foundation, one Diversity Project representative reportedly told participants that researchers were discovering what Native Americans already knew: we are all one (Interview G). Far from building support for the Project, these statements only aggravated concerns. For many, they denoted a lack of awareness of the role that previous scientific claims about a "common stock" had played in legitimating "civilizing" projects that had stripped tribes in the United States of their languages and cultures (Weaver 1997, 16).

Not only were those asked to review the Project not interested in and offended by genetic analyses of their origins, they also feared these studies would threaten their rights to territory, resources, and self-determination. As a *Primer and Resource Guide* on genetics published by the Indigenous Peoples Council On Biocolonialism (IPCB) later explained:

> Scientists expect to reconstruct the history of the world's populations by studying genetic variation to determine patterns of human migration. In North America, this research is focused on validation of the Bering Strait theory. It is possible these new "scientific findings" concerning our origins can be used to challenge aboriginal rights to territory, resources and self-determination. Indeed, many governments have sanctioned the use of genomic archetypes to help resolve land conflicts and ancestral ownership claims among Tibetans and Chinese, Azeris and Armenians, and Serbs and Croats, as well as those in Poland, Russia, and the Ukraine who claim German citizenship on the grounds that they are ethnic Germans. (Indigenous Peoples Council On Biocolonialism 1999, 24)[228]

These fears were not based in fantasy. For example, some tribes in the United States grew very concerned about a bill introduced into the Vermont state legislature (Vermont H. 809) proposing that the state "establish standards and procedures for DNA-HLA testing to determine the identity of an individual as a Native American, at the request and expense of the individual" (Yona 2000).[229]

Finally, others worried that in addition to eroding legal rights, genetic studies of origins might threaten practices that sustain and create tribal identities, such as writing and telling stories. As the Kiowa novelist Scott Momaday explains, telling stories is

> imaginative and creative in nature. It is an act by which man strives to realize his capacity for wonder, meaning and delight. It is also a process in which man invests and preserves himself in the context of ideas. Man tells stories in order to understand his experience, whatever it may be. The possibilities of storytelling are precisely those of understanding the human experience. (Vizenor 1978, 4)

Some asked to comment on the Diversity Project at the MacArthur meeting worried that human genetic diversity research threatened to place the power to tell these life-sustaining stories in the hands of scientific and technical elites who were almost exclusively non-Native. This shift in power, one participant reportedly argued, might in a very real way amount to genocide.

Given these historically embedded problems with scientific efforts to study and "discover" the history and origins of Native peoples, many of the MacArthur meeting participants had difficulty understanding why Diversity Project organizers thought their Project would be of interest to their communities.[230] Not only could most not understand this suggestion from their own perspectives, they could not understand it from the perspective of what organizers had stated about the Diversity Project's goals and interests. For example, in conversations at the MacArthur meeting, and in documents distributed beforehand, organizers only presented research questions defined by the scientists who had proposed the initiative. How, one MacArthur participant reportedly asked, did organizers reconcile this difference between the claim that the Project would answer community questions and the reality that the only questions articulated so far had been those of scientists?[231] NAmC representatives responded by arguing that the Project was changing and now sought to address issues of interest to populations. They cited health-related issues as the primary example.[232] One member of the NAmC, for example, talked about the improvements that population sampling could bring to studies of diabetes. Others discussed the importance of assessing natural variation for studies of disease. And, indeed, some at the meeting did express interests in these studies, and would continue to be interested in the possible health benefits. But the question remained: would the research proposed by Diversity Project organizers lead to such benefits?

As future experiences would make clear, many had reason to be suspicious. In 1997 a researcher contacted a tribe about the possibility of participating in a diabetes study. Further inquiries into the request revealed that no such study existed (Interview G). Instead, researchers sought to collect tissue samples for a biological diversity preservation project. In light of such experiences, some began to ask whether researchers were interested in their health, or whether they simply wanted their DNA for their biological preservation projects. They began to refer to these promises of health benefits as false hopes that had been tacked onto the Diversity Project goals in order to attract populations to the Project (Interview G; Montana field notes, 10 October 1998).[233]

Indeed, even some organizers of the Diversity Project recognized that disease research had been added to entice groups to participate, and that the promise of health benefits had perhaps been disingenuous. As one organizer explained the problem:

> At that point [after the Bethesda ethics meeting], you see, the issue really started to come up, a serious issue: Is it an evolutionary study, or in the long run would it have something to do with medicine? . . . If you want to study evolution you don't have to study families. . . . sample sizes are smaller, you are not interested in association between diseases and markers, so you don't need so many data points. So I would have to say frankly that there was a conflict within the Project about that. Because you are trying to sell it on its political utility, but the sample sizes you could eventually afford to pay for weren't going to be big enough to do genotype/phenotype [studies]. (Interview A4)

In other words, although organizers thought disease research would be important because it would give something back to communities, they also later recognized that it moved beyond the scope of their proposed research. Such research would require genotype/phenotype information that the Project did not plan to collect, and money it never expected to secure.[234]

As this case of health research illustrates, although Project organizers might sincerely have wanted to ask questions of interest to the populations it proposed to sample, it remained a reality that these populations had not proposed the Diversity Project; population geneticists and molecular evolutionary biologists had. Thus, not surprisingly, the questions the Project could support reflected the interests of the originating scientists.

Material Practice

To change human genetic diversity research so that it reflected the interests of populations as well as researchers, many Diversity Project organizers believed that members of populations should become involved in the discussions and meetings where research agendas took shape. However, in practice, these discussions and meetings continued to embody the values of the scientists and ethicists who facilitated their projects. Rather than opening up opportunities for change, they threatened to merely reproduce patterns of exclusion and privilege. The following account of three meetings held on human genetic diversity research between the fall of 1996 and the fall of 1998 illustrates this problem.

OPEN DOORS? MONTREAL, SEPTEMBER 1996

On 6 September 1996, the First International Conference on DNA Sampling was convened at the Delta Hotel in Montreal, Canada. As an organizer of the meeting later recalled: "The purpose of the meeting was to find out on DNA sampling who is doing what. Who is sampling for epidemiology? Who is sampling for newborns? Who is sampling for diversity?" (Interview H). Reminiscent of the claims made by organizers of the 1992 Diversity Project anthropology workshop at Penn State, organizers of the Montreal conference argued that they sought to gather the facts.[235] Who was doing DNA sampling, and for what purposes? Meetings discussing issues (like commercialization) would follow. All were "welcome to attend," but it would be the scientists—those who allegedly knew "the facts" about human genetic diversity research—who would speak.

This, however, was not the experience of the members of indigenous and public interest groups who attempted to attend the conference. Coincidentally, on the day that the DNA Sampling conference began, a meeting of the scientific and technical advisory committee to the United Nations Convention on Biological Diversity (CBD) adjourned just a few blocks down the street. Members of indigenous and public interest groups had been asked to speak at this CBD meeting, but the same invitation had not been extended at the conference up the street on DNA sampling—an issue of equal or even greater concern to indigenous peoples. Although not explicitly invited, members of these NGOs did try to attend the DNA conference in order to voice their concerns.[236] However, security guards blocked their entrance. Once rebuffed at the door by not only hotel security guards, but the Royal Canadian Mounted Police and the Montreal police, members of the Assembly of First Nations, Cultural Survival Canada, the Asian Indigenous Peoples' Pact, the Asian Indigenous Women's Network, RAFI, and numerous other indigenous and public interest groups set up a protest outside of the Delta Hotel challenging conference participants to " 'check their ethics at the door' " (Benjamin 1996).

The conference organizer reportedly defended her decision to not invite indigenous people to speak on the grounds that the issues surrounding DNA sampling were no more relevant to any one population than another, and it simply was not possible to invite all the potentially interested groups (Interview H). She did, however, insist that all had been welcome to attend, and could have exercise their democratic right to speak form the conference floor (Benjamin 1996). To make participation easier, organizers offered a special reduced student conference rate of $35 (Interview H).[237]

This defense did not ring true with the CBD representatives who experienced something very different on the ground. First, not only did they believe that they had been denied their democratic right to speak from the floor, they reported that they had been denied their "much less democratic right" to pay admission to attend the conference, as they had been removed by the police before they could even reach the registration desk (Benjamin 1996). Second, to claim that representatives of indigenous groups had not been invited to speak at the meeting because it was a "fact-finding" meeting—and so would not address "issues"—represented a blatant rewriting of history. If commercialization was one of these issues—as the organizer claimed—then the DNA Sampling meeting did address issues. Indeed, commercialization was the topic of Plenary 6, Commercialization and Patents ("First International").

In short, representatives of indigenous and public interest groups who attempted to attend the DNA Sampling meeting in Montreal found the organizers' defenses of the meeting structure weak and only demonstrative of how indigenous people were being excluded from the debate about human DNA sampling and human genetic diversity research. As their Open Letter to the meeting charged:

> This conference could have provided an opportunity for the open debate and public scrutiny that is so desperately needed. Instead, the indigenous peoples who are the primary targets of DNA sampling have been largely shut out of the conference and *participation* restricted to the proponents of mass sampling and commercialization of human genes. As a consequence, the "First International Conference on DNA Sampling" appears likely to stand as merely the most recent example of the genomic industry's callous disregard for the rights and safety of the peoples it targets [italics added]. (Cultural Survival Canada et al, 1996).

Although claiming in words to be participatory, critics argued that organizers of conferences on human genetic diversity research had acted to exclude the very groups who would be the objects of research.

SHOULD WE SPEAK? PALO ALTO, CALIFORNIA, JANUARY 1998

Charges that the debate about the human genetic diversity research—in particular, the debate about the Diversity Project—excluded Native and indigenous voices reached a crescendo in January 1998. In that month, Diversity Project organizers invited members of American Indian tribes in the United States and Canadian First Nations to gather at Stanford University to discuss the Project. This was the first time that organizers of the Diversity Project attempted to use their MacArthur Foundation money to support a meeting between Project organizers and Native peoples since the March 1994 MacArthur meeting in San Francisco.[238]

Diversity Project organizers placed Dukepoo in charge of organizing the meeting and agreed to provide funds for a separate planning meeting for those invited to attend. Dukepoo argued that this would enable those working on human genetic diversity research in their communities to meet with others and share their knowledge and experiences. The NAmC committee agreed to this stipulation.

Reportedly, however, members of the Diversity Project's North American Regional Committee set the agenda for the main meeting. This agenda consisted of gaining input and feedback on the NAmC's Model Ethical Protocol. Members of the NAmC had worked for several years creating the Protocol largely with the hope that it would provide ethical guidelines that would respond to the concerns of the indigenous groups it hoped to sample. Project organizers now asked members of these groups to engage with this document.

However, much to the organizers' dismay, this discussion never really got started.[239] Instead, the meeting quickly turned to concerns about the Project's goals. Just months before, a National Research Council committee released a report that stated that "there was no sharply defined proposal [for the Diversity Project] that it could evaluate" (National Research Council 1997, viii). Invited participants at the Stanford meeting adopted the same criticism.[240] Before discussing the proposed Model Ethical Protocol, many wanted to know exactly what the Project proposed to do, and what benefits from it would accrue to their communities. Since members of the NAmC failed to convey any clear benefits, those gathered refused to discuss the Project further. The meeting ended, Dukepoo reported, in an "impasse because of unclear goals" (Dukepoo 1998b).

THE RIGHT TO FRAME THE DEBATE: FLATHEAD RESERVATION, MONTANA, OCTOBER 1998

Following this meeting, some in attendance decided that they needed to organize their own meeting on their own territories to discuss the Diversity Project. Organizers of this follow-up meeting, which took place in October 1998 on the Flathead Reservation in Montana, invited members of the Diversity Project's NAmC to attend. However, the NAmC refused the invitation on the grounds that the meeting was "biased." In particular, the then chair objected to the use of the words "colonialism" and "biopiracy" in the title of the meeting: Genetic Research and Native Peoples: Colonialism through Biopiracy. Judy Gobert, dean of Math and Sciences at Salish Kootenai College, co-organizer of the meeting, and an attendee at the Stanford meeting, responded:

> As to the conference theme "Colonialism through Biopiracy," I can do nothing to change it. I strongly suggested to Tribal Leadership that we choose a different theme. They were very adamant about retaining this theme. Their position has been and continues to be that HGDP, the

> Env. [Environmental] Genome Project and others must provide more than a "for the good of the world" or the "pursuit of knowledge" as a rationale for Tribal people to participate in such studies. Considering these are the very people you want to collect samples from, I wouldn't argue semantics with them. Tribes feel very strongly they have been exploited from the time of the first landings on this shore. Everyone who knows of these projects sees them as benefiting the researchers, not Tribal peoples. If you are afraid to argue your points before Tribal Leaders, by all means, don't come. If your project is a way to "eliminate racism" and you feel strongly about the validity of your research, argue your position. (Gobert 1998)

Although representatives of NIH's Environmental Genome Project, as well as other program directors at the NIH involved in human genetic variation research, did attend, no representative of the Diversity Project took part in the Montana meeting. Project organizers, as in the past, refused to recognize that their Project might be linked to the histories of both racism *and* anti-racism.

Experiences at these three meetings made it clear to those recommended for sampling that researchers and bioethicists had in many ways already set the terms of their participation in the dialogue. Accordingly, many viewed with great skepticism offers to enter into partnerships with Diversity Project organizers.

Participation: A Backdoor to Objectification and Co-optation?

Those targeted for sampling worried not only that a level playing field did not exist that would facilitate meaningful democratic participation, but also that participation under the existing rules of the game would lead to objectification and cooptation. As a Native attorney in health policy explained, the image he got from the entire Model Ethical Protocol (and not just from the Partnerships section) was not one of empowered subjects, but of reactive objects:

> It makes tribal people look absolutely unsophisticated. I get images of someone who really has not had much contact. It gave me the image of tribal people as object. . . . it did not give me much image of them as people. They were objects. They were reactive objects, however. And the purpose of this—one of the things the scientists would have to do—is to elicit the proper reaction. (Interview F)

Indeed, the words "scientifically unsophisticated" do appear on the very first page of the Model Ethical Protocol. Specifically, it states: "[This Protocol] deals expressly with the ethical and legal issues that are raised when a project seeks DNA explicitly from populations, not individuals, especially when those populations maybe be *scientifically unsophisticated* and politically vulnerable [italics added]" (North American Regional Committee 1997, 1). Again, drafters of the proposed Protocol assumed that populations, and not researchers, lacked necessary knowledge and sophistication. As we have already seen (in this chapter and chapter 5), Diversity Project organizers argued that members of groups would see the value of the research and would be motivated to formulate their own research questions, if given the opportunity to learn about the Project and gain some "sophistication" on genetic matters.

Some, however, saw another side to these educational "opportunities." Instead of honest efforts to address community needs, they feared that they would become bribes to encourage participation. Indeed, professors at tribal colleges did report cases where researchers offered scholarships to students who would provide genetic samples (Shelton 1998, 7). An organizer of the Montana meeting described it this way:

> Well, there is certainly nothing wrong with having more Native scientists, but let them determine the areas of interest from a community perspective and from the basis of community needs. What they are talking about when they say that [partnership] is "let's pull off the few Indian students to work with us so that we will have access to their DNA and their communities and what they will learn is what we're doing in the process." . . . It's not "let's sit down with Indian scientists and upcoming Indian scientists and let's look at [what] you know"— You could go a step further here and look at how many Native students have access to the sciences are being nurtured to become scientists. Who is funding it and how much is being put into [it] so that you can have scientists who sit there and develop their own topics based on their own needs and their community needs and interests? *What they are trying to do is to get participation into their own agenda* [italics added]. (Interview J)

In other words, the problem was not with participation per se, but with the kind of participation proposed by Diversity Project organizers. This kind of participation did not provide a structure in which the interests of those sampled could be expressed.

These criticisms help to explain why neither education nor participation in genetic projects could be taken as universal goods. Rather, different programs for education and different models for studying human genetic variation connected to different relations of power—some empowering and some disempowering. As the sociologist of science Steven Epstein has noted in his work on the democratic potential of AIDS activism, those models of knowledge and power that seek to "simply disseminat[e] scientific knowledge in a 'downward' direction—creating a community-based expertise—seem potentially naïve, or at a minimum, insufficient. In the worst-case scenario, such a strategy transforms the recipient of knowledge into an object of power" (Epstein 1991, 55). This was precisely the concern of many asked to review the Diversity Project: far from permitting them to function as active subjects, the Project would further objectify and disempower sampled groups.

Offers to provide education would continue to prove a very difficult matter to navigate. As one member of an indigenous rights organization who had been working at the UN level on these issues put it: "What are you going to do? Say education is bad, that people should not be informed? [pause] It's the liberal idea of education" (Interview K).

In addition to these worries about who defined and controlled the flow of knowledge, some worried that participation threatened to drain valuable resources that could be used to address more immediate problems. As made clear by the opening words of the chair of the Confederated Salish and Kootenai Tribes at the October 1998 meeting in Montana, many people could not be at the meeting because they were in Washington to argue for water rights, and so could not come and talk about genetics (Montana fieldnotes, 10 October 1998). And as an attendee of the 1998 Stanford meeting further explained:

> Just in December, elders froze in South Dakota because of substandard housing. And then you look at the millions and millions of dollars that are being spent to research indigenous peoples' DNA with absolutely no regard for our lives, and maybe even no intent to ensure our survival as peoples. I honestly don't think that there is much of an interest really to make sure that we survive as who we are. (Interview J)

Sampling indigenous DNA might answer questions of interest to scientists, but would it contribute to the survival of an oppressed people? Would genetic research lead to greater awareness of and attention to the health needs of research subjects, or would it merely drain resources from basic health care?[241] Those asked to review the Project asked organizers to answer these basic questions.

Some also noted that doing the work needed to participate in the debates placed a burden on the groups targeted for sampling. As one indigenous rights leader explained, there were many demands on his time. Whose time and money, he asked, would be spent raising awareness about these issues? (Interview K). And as Frank Dukepoo told the National Bioethics Advisory Commission: "Indian people have turned to me to do a lot of interpreting of what's going on here. We don't have the person power to deal with these issues" (Dukepoo 1998b). In short, many asked to participate in the dialogue about the Diversity Project argued that it placed an undue burden on them to learn about and raise awareness about research that did not support their interests.[242] Some clarified that they were learning about the Project not because they thought it could help communities, but because they feared it would endanger them:

> The reason why I work on it now is because I understand from walking in two worlds what one world is trying to do and the potential dangers of that to the other world. So I am compelled morally to work on this stuff now. And it has become a serious threat. In my estimation, this could happen. (Interview F)

For many years, this point proved difficult for Project organizers to understand. They believed with sincerity that the Project would interest and benefit all people.[243] Many members of the populations they wished to sample remained unconvinced.

PARTICIPATION: THE STRATEGIC CONSTRUCTION OF VOICES IN SOCIETY?

To many potential research subjects, it seemed that Project organizers proposed partnerships mainly for reasons of political expediency. In an age of identity politics, finding ways to permit the participation of minority ethnic groups has emerged as a prominent strategy of legitimation (Amit-Talai and Knowles 1996). Liberal democratic states, such as Canada and the United States, have created numerous institutional committees, tasks forces, hearings, and programs designed to increase opportunities for minority representation. However, instead of creating a more representative democracy, critics argue that such entities only serve the state's interest in creating "institutionalized exhibitions of representation" (Amit-Talai 1996, 99).

To understand how this interest in creating racial- and ethnic-group participation affected the Diversity Project, one need only recall the MacArthur Foundation's stipulation that it would only fund the Project's effort to write an ethical protocol if Project organizers consulted with Native and indigenous communities.[244] In this and other instances it became clear that Project organizers would need to establish that they had listened to "the voices" of the groups their research wished to sample.

This, however, raised complex questions about the identity of groups, and who could speak for them. These questions emerged soon after Project organizers began trying to consult with groups. Those who attended the January 1998 Stanford meeting, for example, emphasized that they did not speak for anyone but themselves, except on the rare occasion when a member of a tribe had received authorization from his or her tribal council to speak for the tribe. Indeed, at Stanford reportedly only one person had received this kind of authorization (Interview G).

Yet whether viewed as individuals or as official tribal spokespersons, those who attended meetings realized that they ran the risk of becoming token voices. As one participant at the Stanford meeting explained: "One of the problems I would say is tokenism. It's easy to say that we talked with our Indians and so now we are moving forward with our stuff. The question is what do you mean by talk to and what authority do they have to speak for anyone?" (Interview J). Or as an indigenous rights leader observed, if you talk to them, "by osmosis you help [the Project]" (Interview K). The problem of tokenism presented those asked to represent groups with a dilemma: one wanted to find out what was happening with the Project, but to enter into dialogue with its organizers was to risk lending the Project legitimacy. To handle this problem, one indigenous rights leader reported that on occasion he would accept invitations to lunch with Diversity Project organizers. He tried to be in touch only enough to stay abreast of what was happening.

A Native lawyer described a similar dilemma that he faced upon being asked to review the Model Ethical Protocol: "Why should I go in and critique section by section when you don't even have the big picture right here? . . . I grant a type of legitimacy if I start critiquing it section by section." (Interview F) Another participant in the debate explained why they would not discuss their concerns about partnerships with Project organizers:

> [B]ecause then you are saying, well, essentially first we agree that what you are doing is good and let's make sure we get some students trained in the process. So, no, I have not said that, and I wouldn't make them an offer either because I think it would be a tool to co-opt. It definitely provides access into communities. (Interview J)

In short, the problem was not that Diversity Project organizers failed to invite members of groups to participate. Quite to the contrary, organizers tried very hard to include them as they attempted to gain the legitimation that came with consultation.[245] The problem was that this effort threatened to produce "Native American" voices in a manner that would, in the view of the concerned groups themselves, only reproduce structures of objectification and co-optation.

INTEGRATION OR SEPARATISM: STRUGGLES OVER THE CREATION OF AN AFRICAN AMERICAN VOICE

Members of tribes in the United States and First Nations in Canada were not alone in their worries about Project organizers' constructions of voice. Like those asked to speak for Native communities, some of those asked to speak for African Americans feared that organizers would strategically construct an African-American voice of support for the Project. This concern became public in January 1994 when a consortium of African-American social and biological scientists met in Washington, D.C., to draft the *Manifesto on Genomic Studies among African Americans*. Among other things, these researchers sought to respond to a variety of negative experiences with the Project, including an episode that transpired at the 1993 World Council of Indigenous Peoples meeting in Guatemala.[246] At that meeting, the chair of the NAmC ethics subcommittee, Henry Greely, reportedly stated that African Americans supported the Project.

This claim alarmed an African American anthropologist who also attended the WCIP meeting. As explained above, organizers did not originally propose to sample African Americans. Further, although organizers did add Georgia Dunston to the group of scientists invited to planning meetings, and eventually appointed her a member of the North American Regional Committee, at the time of the WCIP meeting many African American biological anthropologists and geneticists still felt that most Diversity Project proponents opposed the inclusion of African Americans on the grounds that they were not a meaningful biological group for the purposes of the Project. Given this, the African American anthropologist who attended the WCIP meeting interpreted Greely's statement that African Americans supported the Project as nothing more than a political statement designed to generate the support of racial minorities and indigenous people. She reasoned that if organizers could claim that the initiative had the support of African Americans, a group perceived by many outside the United States as occupying an adversarial relationship to the dominant power structures, then it might become more palatable to indigenous groups (Interview 1999).

In response to this reported appropriation of race as a political instrument, and the absence of African Americans in the database of the Human Genome Project, African American biological and social scientists wrote the *Manifesto*. In it they explicitly cited the "misrepresentation of African American interests regarding genomics studies by Hank Greely (Stanford University) . . . at the World Council meetings in Quetzeltenango" as a reason for concern that African Americans would not be fairly represented in the genomics revolution. As a response to past problems, the

Manifesto called for the "full inclusion" of African Americans "in any world survey of human genomic diversity." Clearly referring to the Diversity Project, it proclaimed: "The inclusion of African Americans is not optional in any world survey, particularly if US taxpayer monies are the funding source for such efforts" (Jackson 1998, 165).

These claims and counterclaims about the interests and views of Native peoples and African Americans illustrate some of the fundamental issues of co-production at stake in the seemingly beneficent effort to facilitate the participation of these groups in the Diversity Project. As in the case of group consent, this effort raised complex questions about the role the Project would play in constructing group identities and voices in society as it attempted to understand human biological diversity. In multiple contexts, organizers drew upon Dunston's G-RAP project and Blakey's African Burial Ground Project to argue that both African Americans and people more generally were interested in human genetic diversity research. However, these claims raised questions about who, if anyone, should constitute a group labeled African American, who could speak for this group, and how a voice for this group might be employed to legitimate research. These questions about the construction of groupness and spokesmanship have long and embattled histories in the United States and globally (Churchill 1994, Amit-Talai and Knowles 1996, Alvarez 1997, Barrig 1999).[247] The Partnership provision of the Model Ethical Protocol faltered as it failed to address these fundamental questions about the constitution of the social order embedded in Diversity Project organizers' efforts to order the diversity of nature.

PARTICIPATION: THE INADVERTENT CONSTRUCTION OF GROUPS IN NATURE?

Connected to these questions about how to establish groups with voices and interests in society was an equally complex set of questions about how to establish the existence of groups in nature. Recall that the original proposal for the Project identified "isolated human populations" as those populations of greatest interest to the Project (Cavalli-Sforza et al., 1991, 490). As proposers of the Project explained in *Genomics* in 1991: "It is *only* from knowledge of the gene pools of *these* populations that we can hope to reconstruct the history of the human past" [italics added] (Cavalli-Sforza and Bowcock 1991, 495). These "isolated human populations" were contrasted to "populations of relatively recently mixed ancestry such as those in most urban regions of the world" (Cavalli-Sforza et al., 1991b). Such mixed populations contained much less information about early human history and evolution. African Americans clearly fell into this less informative category.

By 1993, however, Project documents began to state that the Project did not, as the original articles in *Genomics* clearly stated, seek to merely sample the genomes of "aboriginal populations" that were "rapidly disappearing" (Cavalli-Sforza and Bowcock 1991, 495). For example, in a document written in 1994 entitled "Answers to Frequently Asked Questions about the Human Genome Diversity Project," members of the North American Regional Committee asserted: "This Project is not an effort to collect samples from isolated populations in danger of disappearing. It intends to take a representative sample of *all human populations*, including those in Europe and North America. No group is necessarily excluded" [emphasis added] (North American Regional Committee 1994c, 6). Further, as documented in the first part of this chapter, the incoming chair of the NAmC argued that African Americans would not be excluded from the Project, as some African American physical and biological anthropologists feared (Weiss 1993).

This transformation in ideas about inclusion set forth in Project documents accompanied a change in notions of inclusion occurring at the same time at the National Institutes of Health (Epstein, forthcoming, 7). In 1993 President Clinton signed the NIH Revitalization Act of 1993, a piece of legislation that contained within it provisions calling for the NIH to include "women and members of racial and ethnic minority groups" in clinical studies (U.S. Congress 1993, Epstein, forthcoming). Supporters of these provisions argued that they heralded the end of privileging the health research needs of white men. Analogous to claims made by Diversity Project organizers, these reformers argued that to date most biomedical research had been conducted on white males, creating a Eurocentric bias.

In the case of both the Diversity Project and NIH Revitalization Act, claims about the need to include racial and ethnic groups in research proved controversial.[248] At the same time that some leaders of the Diversity Project and some supporters of the NIH Revitalization Act called for the inclusion of race as a variable, other Diversity Project leaders and other public health officials in minority communities called for the end of the use of race. For example, Harold Freeman, the head surgeon of Harlem Hospital, explained in a January 1998 letter to President Clinton that race has no basis in biology and should be abandoned as a variable in biomedical research (Freeman 1998). In the same year a manifesto appeared in the *American Journal of Public Health* calling for the end of the use of race in public health research (Fullilove 1998). These claims resonated with those of Luca Cavalli-Sforza who, in his 1994 address to a special meeting of UNESCO, had argued for the abandonment of the category of race and its replacement by the more scientifically precise term population (Cavalli-Sforza 1994, 11). Additionally, in *The History and Geography of Human Genes*, a volume published in the same year, Cavalli-Sforza and his colleagues labeled "the concept of human races" a "scientific failure" (Cavalli-Sforza et al. 1994, 19). As explained in chapter 3, scientists like Cavalli-Sforza employed multiple, seemingly contradictory concepts of race. Calls to use racial categories to organize the Diversity Project raised more fundamental and unresolved questions about which concepts of race, if any, could have utility for biological and biomedical research.[249]

Some worried that by failing to acknowledge these questions, the Diversity Project would only serve to reduce biological diversity to political diversity in a manner that would produce meaningless research. As one European leader of the Project explained, placing a focus on sampling major ethnic groups in the United States would prevent answering questions of interest to population geneticists:

> I mean, the difficulty with the U.S., the only indigenous populations you have got are the American Indians. They are actually not terribly interesting if you want to study most of the world. It's a rather, it's a very interesting area, but it's a rather recondite area if you really want to know what goes on over the rest of the world. If you want to know that, the American population is the worst in the world to study. The only way to understand the Americans is to understand everywhere from where they came, because they are such a mixture. So, how are you going to have American studies then? If the Americans want to understand themselves, they have got to go abroad. So how are they only going to fund studies done in America for samples that are collected abroad? It is not easy to rationalize that in my view, and it reflects one of the problems of the attitudes, the ways that things are looked at there. (Interview A)

For this scientist, the criteria that populations be indigenous remained central. Given this, American Indians were the only populations of interest in the United States. A study of these populations would be interesting, he argued, but would be of limited value to understanding the evolution of the human species throughout the world—the Diversity Project's true goal, from his point of view. Advocates of such research, he contended, were looking at things through a politically correct (and thus scientifically incorrect) lens.

Conclusion

Given new demands and government policies for the inclusion of women and minorities in biomedical research in the late 1980s and early 1990s, it is perhaps not surprising that some proponents of the Diversity Project argued for the inclusion of major ethnic groups as both research subjects and active, knowledgeable participants. However, as this chapter illustrates, participation proved not to be such a simple and objective good as organizers had imagined. It could not be transported like a ready-made instrument to address the concerns of the diverse groups the Project hoped to sample. Instead, Diversity Project organizers' efforts to create a project that conformed to affirmative action policies only created new concerns.

These concerns connected to fundamental political and conceptual issues embedded in the long history of efforts to employ racial categories for various human purposes. Project organizers' alleged use of the participation of one individual African American geneticist to demonstrate the interest of the group African Americans sparked long-standing fears and questions about tokenism. Their attempts to constitute Native and African American voices for the purposes of Diversity Project research tapped into deep worries about the appropriation of race for projects that in effect drained important resources away from addressing the basic needs of communities. These concerns—conventionally understood as political—connected to conceptual questions about the construction of racial categories for the purposes of representing nature. In particular, some population geneticists worried that U.S. racial and ethnic groups defined on the basis of societal norms and practices, such as African Americans, did not exist in nature as natural kinds. Thus, sampling these groups in the United States would do little to advance the goals of the Project.

Project proponents' procedural choices did not provide the tools needed to address these complexities. These choices were built around frameworks that separated science from politics. Within these frameworks it made sense to attempt to resolve questions about who should participate in the Diversity Project debates, and how, by appealing to what they viewed as scientific practices on the one hand (e.g., collecting "the facts"), and political practices on the other (e.g., affirmative action policies). However, in the case of human genetic variation research, a domain where the connections between science and politics are impossible to deny, these two domains proved inseparable. The different histories of racism contained within policies and practices of inclusion in American society could not be disentangled from the different histories of racism embedded in ideas and practices of inclusion in human population genetics research. Focused on the instrumental value of what they perceived as already formed normative practices and tools, proponents were constantly caught off guard as members of the indigenous populations they hoped to sample repeatedly rebuffed their offers of participation.

As we will see, the frustration generated by the failure of these sincere efforts extends far beyond the Diversity Project. My argument is not that the Diversity Project uniquely proved unable to create a conceptual and political order that could accommodate human genetic diversity research. To the contrary, problematic questions first raised in the context of debates about this initiative today still confront researchers, policy makers, and potential research subjects. Now is an important time to reflect on what these earlier struggles can reveal about current dilemmas generated by efforts to use genomic tools to understand human diversity.

Chapter 7
Conclusion

The Diversity Project today consists largely of unimplemented texts, and work on the initiative moves forward only slowly with limited resources from the French Centre pour l'Etude du Polymorphism Humaine (CEPH).[250] Yet despite its failure to gain widespread support, the Project continues to be extremely important for understanding the future of biology. Today, numerous studies of genetic diversity in the human species are pushing ahead under different names, and with different goals and ideals.[251] Proponents of these new efforts have attempted to distinguish themselves from Diversity Project organizers by arguing that their initiatives will answer more profitable questions about human health, and either avoid, or adequately address, ethical problems. Yet, experience indicates that promises to overcome the problems raised by the Diversity Project are likely to be overstated.[252] The refusal of the National Congress of American Indians to give the National Human Genome Research Institute consent to use samples collected from tribes in the United States in their DNA Polymorphism Discovery Resource (PDR) provides only one example among many of continuing trouble (Interview S).[253] In this case, the NHGRI ultimately decided to strip racial and ethnic markers from all of the samples included in the PDR, arguing that it would "take time" to resolve the ethical problems associated with the use of population-specific information (Collins et al., 1998).[254] This decision upset many scientists, who argued that without racial and ethnic information the samples would be largely useless (Greely 2001, Interview A; Interview C). The dispute speaks volumes about the continued importance of racial categories as tools in human genetics, and the unresolved controversies surrounding their use.

These and other episodes suggest that the debates about race, identity, and governance raised by the Diversity Project will only deepen in importance as human genetic variation research rises on the agendas of private and public research institutions around the globe. Far from allowing us to move beyond race, as many have hoped, these studies instead only highlight the centrality of racial categories in the human sciences. Debates over their meaning and proper definition will continue to be pivotal to determinations of what defines human identity at the level of the genome, as well as which human groups are entitled to full rights of voice and representation at the societal level.

These debates will remain controversial, and quite properly so, since they arise not from the incompetence of a few unenlightened organizers (an explanation often used to explain the failure of the Diversity Project), but from unresolved questions about what scientists and other members of contemporary cultures and societies want to know about the differences among human beings. Normative questions are bound to arise as people increasingly hang their hopes and ambitions on scientific research—looking to it both to improve the quality of their lives, and to illuminate issues of identity and difference. To be responsive, science and scientists must craft new tools and processes for operationalizing the concept of diversity—with all the dangers that this move entails. In this concluding chapter, I reflect on what the story of the Diversity Project can tell us about this continuing predicament.

Working Worlds Apart in One World

One striking feature of the story of the Diversity Project is the lack of connection between organizers' and critics' understandings of the initiative. The organizers envisioned a humanistic endeavor that would benefit all of humankind. They believed that, far from threatening to fracture the world along racial and ethnic lines, their initiative would serve to fight racism and unite humans across the globe, if only by involving them in a common project of genetic research. To put it mildly, the subsequent charges of racism, colonialism, and vampirism alarmed the well-meaning proposers of the DNA sampling initiative. The early proposers of the Diversity Project assumed that techniques of population genetics provided the most scientifically rigorous way to study human diversity. Charges of drawing upon nineteenth-century racial ideologies produced puzzlement and dismay. The mismatch between organizers' understandings of themselves as good, anti-racist scientists, and critics' charges that they were furthering colonialism and practicing bad science is not unique to the Project. Many well-meaning researchers—such as those at NHGRI—

have been accused in recent years of doing ill when their goal had been to do nothing but good.

I have argued that these divergent perceptions derive partly from the prevalence of analytic frameworks among scientists, policy makers, and ethicists that fail to bring into sharp focus the intertwined epistemic and political contexts in which biological science is now asked to operate. As much empirical work in science and technology studies has demonstrated, the production of knowledge can no longer be understood as a task for a single set of disciplinary experts operating in narrowly defined contexts of scientific research (Gibbons 1994). In the case of the Diversity Project, for example, the proposal to undertake a worldwide survey of human genetic diversity raised scientific questions that were important not only to population geneticists, but also to physical anthropologists. At the same time, it generated politically sensitive questions that were important to the groups targeted for sampling. Organizers of the initiative often too slowly understood these multiple settings in which their Project would take shape. Further, convinced that the world would be a better place if research on human genetic variation flourished, proponents too often assumed that other scientists, and people more generally, would embrace their Project if only they understood its methods and goals.

For the Project to expand, or even survive, a different framework would have been required—one that would have prompted organizers to recognize the limits of their scientific as well as their moral and political perspectives, and one that would have encouraged them to understand how others might assess the meaning and value of the Project they had proposed. As early organizers of the Project learned, not even all scientists interested in human biological diversity agreed that the tools of population genetics would be the primary ones needed to design a survey of human genetic diversity. Many physical anthropologists, for example, believed that tools for understanding culture would be primary. Others asserted that the subjective experience of being a member of a population was indispensable to the Project.

Further, not all scientists, and not even all Diversity Project organizers, shared the same understanding of racial categories. Some believed that scientists could continue to use racial categories as long as they properly limited their use of these categories to biomedical contexts. Others believed that racial categories could be used to define human identity, but only if these uses incorporated an anthropologist's sophisticated understanding of culture. Still others advocated the use of racial categories in both medical and genealogical analysis, but only if supplemented by the subjective knowledge of sampled populations. Finally, some held multiple views. Thus, population geneticists and physical anthropologists who supported the Diversity Project at some moments rejected a particular racial category (like Negroid), and at other times employed the same category to organize their research.[255]

Convinced of the importance and urgency of their initiative, the scientists who proposed the sampling initiative did not in the first instance adequately engage with these disparate perspectives on race and human diversity. Their call for immediate action might have worked when the initiative involved relatively small numbers of researchers with shared disciplinary backgrounds; it did not work in the case of the Diversity Project, where the participation of scientists from different disciplines, and even nations, was required.

To be successful, the Diversity Project would have required in the early stages not only the perspectives and skills of different scientists, but also the viewpoints of others for whom questions of group identity and origins were central. As we have seen, debates about natural (or biological) human difference cannot be disconnected from debates about how to categorize human diversity in society. Despite their best efforts, organizers could not disentangle themselves from these historically entrenched debates and the emotionally and politically charged discourses of population, race, ethnicity, and colonization that structured them. No matter how foreign the ideas of indigenous groups may have seemed to proposers of the Diversity Project, accusations of racism and colonialism by these groups could not be dismissed by merely asserting that the accusations were politically motivated, or by arguing that the Project would fight racism by demonstrating the fallacy of biological notions of race. Rather, to defuse conflict, organizers would have needed to engage more substantively with these critical viewpoints, and with their associated perspectives on human diversity and race. This would have required acknowledging that they had as much to learn from their critics as their critics had to learn from them.

The Problem of Inclusion

Some close to the Project did feel that its organizers made extensive and adequate efforts to engage with these other perspectives (Greely 2001). As described in chapters 4, 5, and 6, the organizers did put considerable energy and resources toward organizing meetings with both other scientists (at the Wenner Gren anthropology meeting in fall of 1993, for example) and members of tribes in the United States and First Nations in Canada (at the MacArthur meetings in San Francisco in spring of 1994 and Palo Alto in winter of 1998, for example). What, then, went wrong? Why did attempts to learn from and respond to different viewpoints fail? Telling the story of the Diversity Project within the idiom of co-production offers a few insights.

First, charges of exclusion cannot be addressed through simple offers of inclusion. Instead, the meaning and value of inclusion must be interrogated. Time and again Project leaders sought to address the accusation that they had excluded relevant groups, such as 'anthropologists,' and 'African Americans', by including them. Clearly, these efforts did not meet the group members' subjective tests of inclusion. As this book illustrates, this is because organizers drew upon frameworks that assumed the prior existence of groups, and bodies of expertise that could define them. Thus, organizers often did not address critical questions about the very definition of groups such as 'African American,' and their status in nature and society. Nor did they consider how the definitions they used might end up reproducing histories of colonialism and racism. Convinced of the beneficence and scientific importance of their Project, organizers often failed to grasp these deeper ontological, historical, and political dimensions of their initiative.

Second, these fundamental questions about the constitution of groups in nature and society raised by the Diversity Project cannot be addressed by adapting the tools of Western biomedical ethics to already formed scientific agendas for examining human differences (Jasanoff 2002). As chapters 5 and 6 demonstrate, to address the concerns of critics, organizers sought to build an ethical foundation for their Project, codified in the proposed Model Ethical Protocol. This effort faltered as it conceived the task of crafting ethical norms and practices as separate from that of formulating scientific norms and practices. Rather than viewing ethics as integral to the conceptualization of research, many participants argued that the science needed to be worked out first; ethical protocols to accommodate the science could follow. Science, for example, could define the subjects from whom informed-consent should be gained. Ethical protocols could then be written to help researchers gain this consent in an ethical manner. Such an approach bypassed questions about the role the informed-consent process itself would play in defining subjects (in this case, groups), questions that proved of enormous importance to many of the indigenous groups in North America and Australia that the Project hoped to sample.

Finally, the conventional battle lines of scientific controversy hamper dialogue. As we also saw in chapters 5 and 6, participants in the debate on the Diversity Project were quickly sorted into rational supporters interested in science, and irrational critics interested in politics.[256] Organizers viewed critics, such as World Council of Indigenous Peoples, as parties motivated by political agendas that had no science behind them. As the organizers of the October 1998 meeting on Genetic Research and Native Peoples discovered, it was difficult to be critical without being perceived as politically motivated, and thus as an enemy of reasoned debate. The resulting asymmetry—organizers' attribution of political content to the discourse of critics and not to their own discourse—often polarized exchanges and preempted needed discussions.

In short, frameworks organizers drew upon to address problems of inclusion raised by their initiative stumbled as they divided the world into already formed collectivities (anthropologists, groups who should be the subjects of consent, African Americans, etc.), and already-formed and distinct domains (culture, biology, ethics, science, rationality, irrationality, etc.). Within these frameworks, it made sense to proponents to either generate support for their initiative by attempting to enroll some of these collectivities and domains, or to protect it by keeping others at bay. It seemed natural, for example, to turn to a domain called ethics to resolve questions about moral order. It in turn made sense (as the case of group consent illustrated) to turn to scientific experts to answer questions about natural order raised by the effort to create moral order.[257] However, by telling the story of the Diversity Project from within the idiom of co-production, it becomes clear why this approach proved problematic. In short, natural and moral order proved inextricably entangled, making it impossible to construct one without implicating the construction of the other. As a result, the problems Diversity Project organizers faced could not be solved through a straightforward process of defining their scientific objects and then creating ethical and political procedures that could support studying them.

Paradoxes for Science

Far from straightforward, this study demonstrates that generating stable solutions to the problems raised by the Diversity Project, and similar efforts, require navigating a set of paradoxes that have to date escaped the critical attention of scientists and policy makers. I conclude by describing these paradoxes, and by providing some modest proposal for how they might be addressed. Three paradoxes are of particular note.

First, phenomena, such as human genetic diversity, cannot emerge as objects of study independent of moral and social decisions about who we are and what we want to know. In other words, objects simply cannot sustain critical scientific scrutiny if there are no stable moral and social orders to "prop" them up (Jasanoff 2004b). Consequently, technoscientific objects do not precede research projects; rather they are crafted in the very process of constructing the normative regimes needed to accommodate them (Rabeharisoa and Callon 2004). They are not preexisting neutral entities, bur rather human achievements that encode particular epistemological and normative commitments (Daston 2000, Goodman 1978, Hacking 1999).

The Diversity Project provides a striking example. To see this, I return for a moment to the case of group consent. In this instance, drafters of the Model Ethical Protocol attempted to use "researchers' sampling criteria" to define in advance the groups who would be the objects of their research, and thus the subjects who should be included in the Project's ethics process. However, Native attorneys in the United States asked to review the Project objected. Researchers' purportedly scientific definitions of objects, they argued, incorporated decisions about group identity of great political and cultural importance to tribes. Critical physical anthropologists also highlighted the consequential decisions that would be made as organizers used purportedly neutral methods to define populations, noting that these efforts would entail epistemological choices about how one could produce knowledge about the human and its possible variations.

Second, as the Diversity Project episodes demonstrate, and as genomic researchers increasingly acknowledge, efforts to abandon racial categories and to reconstitute studies of human biological differences in a socially acceptable way often unwittingly reinscribe race in new ways (Weiss and Kittles, 2003). As in the case of the Diversity Project, these efforts often begin with the assumption that past studies of human genetic variation erred only when they mixed science with ideologically charged concepts like race. The proposed correction then is to replace these politicized categories with those deemed more objective, such as population and deme (Cavalli-Sforza et al., 1994, Graves 2001). What is overlooked in these diagnoses, and what this book has attempted to demonstrate, is that there is no neutral, apolitical ground to which scientists can retreat to conduct their research. All efforts to study human genetic variation necessarily build on a terrain structured by centuries of prior struggles over the meaning and proper use of race—struggles that thoroughly entwine natural and social orders. Questions about the biological meaning of human group categories—such as African American or American Indian or Pima or Yoruban—simply cannot be disentangled from questions about their social meaning and political utility. Thus, rather than building a world free from race, population-based genomic research represents a new chapter in long-standing efforts to define the proper uses and meanings of racial categories.

Third, genetic definitions of human groups—definitions portrayed in much academic writing of the 1990s as potentially racist (Duster 1990, Wright 1990, Lippman 1991, Paul 1995)—have in recent years emerged as powerful resources for building civic identities that enable individuals and groups to make demands against the state (Jasanoff 2004). Thus, African American geneticists, such as Georgia Dunston, pushed to include African Americans in the Diversity Project. And indeed today Dunston is the director of a National Human Genome Center at Howard University, a center founded on the premise that those of African descent—namely, African Americans—possess unique and valuable genomes and should be awarded resources to study them (Pollack 2003).[258] An analogous development can be seen in the area of biomedical research, where activists, politicians and researchers have in the last two decades succeeded in turning 'women,' 'minorities,' and 'gay, lesbian, bisexual, and transgender' into biomedically meaningful categories, and thus categories that can shape the allocation of research resources (Epstein 2003).

These researchers, clinicians, activists, and policy makers currently find themselves in the paradoxical position of crafting identities that might at once liberate and oppress. Despite the positive potential of identities built around the axes of genomics and biomedicine, the Diversity Project debates illustrate that worrisome questions remain. Will these approaches confer benefits only on those who have the resources to learn the technical skills and language needed to act credibly within genomic and biomedical worlds? Will they serve to privilege academic and expert knowledge in ways that could disempower communities? Will they advance the project of freedom and equality, or reinscribe the very structures that produce inequality in the first place?[259] The Diversity Project debates highlight the importance of these questions, questions that continue to be central to debates about how and whether to include groups in the design and regulation of human genetic variation research.

Constructing Institutions and Knowledge in a Time of Emergence

All of these paradoxes point to a need for new institutional and intellectual frameworks that bring into focus, and enable us to ask about, the complex processes through which natural and political order come into being together. Questions about whether and how to define racial and ethnic categories for the purposes of genetic research must be deliberated along with the associated questions about how to define racial and ethnic categories for the purposes of determining access to insurance, jobs, and health care. Questions about the scientific merit of various DNA sampling projects must be asked along with questions about sovereignty and autonomy. In each case, the latter set of questions—conventionally thought of as political—must be regarded as fundamental to the formulation of research agendas.

In reviewing the history of the Diversity Project, the African Burial Ground Project described in chapter 6 emerges as one hopeful example of this kind of integration. The ABGP explicitly recognizes the latent political dimensions of research in genetics and evolutionary biology. This enabled its organizers to create spaces in which questions about what kind of knowledge the project would produce (and how) were asked along with questions about who would control and define the initiative, and with what political and social consequences.

Yet to look to the ABGP as a hopeful example is not to claim that it addresses all the difficult questions raised by human genetic diversity research. It leaves unexamined, for example, critical issues of voice and representation, such as, Who speaks (if anyone or any one group) for the 'African American community' on issues of genetic research? Some might rightly worry that the ABGP can be used to demonstrate that a group called African Americans support genetic studies of human differences. As we have seen, this concern first arose as Diversity Project organizers reportedly pointed to the one African American scientist on the North American Regional Committee as evidence that African Americans as a group supported the Diversity Project. These concerns continue to be of great relevance today as projects like those sponsored by the Howard University genome center renew the debate about who—if anyone—can legitimately represent African Americans on matters genomic.

To address these concerns, the Diversity Project debates suggest that premature identification of spokespersons for the genomes researchers hope to sample should be avoided in favor of processes that permit deliberation on the meaning of fundamental concepts such as race, difference, community, population, representation, and consent. The creation of these more open deliberative spaces will require all concerned to recognize that science is a human activity that raises not just epistemic and ethical issues, but political ones. This reconceptualization will enable debates about diversity research to connect to a wider range of relevant issues and questions. For example, by understanding the Diversity Project as a potential site for the production of social identities, as well as a site of ethical concern, organizers might have more readily recognized that sovereignty, not informed consent, was the critical issue for those asked to represent Native peoples.

Equally, meaningful deliberation will require the further development of critical frameworks that not only bring to light the political character of discourses (in particular, *scientific* discourse) conventionally thought of as neutral (the conventional domain of ideology critiques), but also the epistemic dimensions of discourse conventionally thought of as political. These frameworks would enable critical observers of science to avoid theories of social constructivism that propagate socially determinist positions— theories that too often reduce science to a mere effect of power (Hacking 1999, Jasanoff 2004).[260] This would do much to alleviate the misconception that critics of science are merely interested in politics (or power) and not knowledge or rational inquiry. As we saw in chapters 5 and 6, this misunderstanding often led Diversity Project organizers to dismiss those asking questions about rights of sovereignty. This frustrated many of those raising these critical questions, some of whom had devoted their entire professional lives to science, and most of whom rejected the accusation that they were anti-science.[261] Recovering the questions about the construction of knowledge embedded in seemingly political domains (such as indigenous rights and affirmative action) would prevent such easy rejection of critical voices, and would facilitate discussions of issues, such as the construction and proper use of racial categories, that reach across communities interested in population genetics, sovereignty, racism, and North/South relations. This book serves as one example of the kind of analysis I am advocating.

Early in the planning of the Diversity Project, organizers warned that if the Project did not succeed, human genetic variation research would continue, but in a less organized and public fashion. Their prediction appears to have come true. Much genomic research on human diversity moves forward today in private companies or in an ad hoc way, accompanied by little or no debate.[262] This should be cause for great concern. For planning and implementation of human genetic variation research to proceed without a broad-ranging societal debate about race, identity, and authority is to build both a less reflective science and a less sensitive society. More specifically, to conduct this kind of research without explicitly considering these issues in the early stages threatens to reproduce structures of race and racism that we seek to transform and overcome in domains other than science.

To be sure, some public efforts and debates do continue even in the absence of a thriving Diversity Project. In the last four years, for example, the NIH has devoted increasing energy and resources to creating a public discussion about how the Institutes will study human genetic variation.[263] As part of their effort, in fall 2000 leaders at the NIH organized a "community consultation" with members of "identified populations" targeted for genetic sampling.[264] This event led to direct policy changes, most notably the requirement that "community consultations" accompany all collections of samples from "identified populations" to be included in the National Institute of General Medical Science's (NIGMS) Human Genetic Cell Repository (Greenberg 2000).[265]

These efforts, however, fall short on several counts. First, they have suffered from a tendency toward institutional caution and tight control. Program officers largely have retained the final word on who would be invited to "community consultations" (Interview X). Second, program officers at NHGRI and NIGMS, and external advisors selected by these officers, created the list of people considered, raising questions about the ability of experts in genetics and Western biomedical ethics to address the broader political questions about representation and identity at stake in defining who will be invited to consultations (a problem described in chapter 5). Third, those involved in the planning of the Diversity Project often were not invited. In the wake of the controversies surrounding the initiative, every effort has been made to contain and carefully manage NIH efforts to study human genetic variation so as to avoid what has been widely perceived as the political troubles of the Diversity Project.

My hope is that the preceding chapters have helped draw into view the problems raised by the very attempt to keep issues conventionally understood as political outside of spaces conventionally thought of as scientific. As this book has demonstrated, these issues are intimately entangled, indeed inseparable. In the light of this demonstration, we might better understand the Human Genome Diversity Project not as an anomaly or as a failure that we can prevent from repeating itself, but rather as a project struggling to do the difficult work that will be required by any effort to study human genetic differences. To the extent that this admittedly painful process leads to a more reflective understanding of the complexities of race, identity, and human diversity, it might even be thought of as a success.

Appendix A
Methodological Appendix

This book draws upon many different kinds of sources, including ethnographic fieldwork, semi-structured interviews, documents produced in the course of efforts to organize a Human Genome Diversity Project, and a variety of published materials, such as critiques of the Project, historical documents and secondary literature, and electronic material (e.g., e-mail lists and news nets). Below I briefly describe these data and how I gathered them.

Ethnographic Fieldwork

From 1995 to 2000 I conducted a multi-sited ethnography of the Diversity Project debates.[266] As opposed to traditional ethnography—where the location of research is determined in advance—in a multi-sited ethnography the "field" of fieldwork emerges as the ethnographer follows complex cultural phenomena in different settings (Marcus 1998). My research required a multi-site approach, because the controversy about the Diversity Project emerged in diverse social worlds. Only by tracing the Project through various sites and communities could I make sense of the complex debates that pervaded and surrounded it.

To follow the controversy, I traveled to many sites conventionally thought of as "worlds apart" (Marcus 1998). They included population genetics and molecular anthropology laboratories at U.S. universities; American Indian reservations in the American West; the Montreal office of the Rural Advancement Foundation International (now called the Action Group on Erosion, Technology and Concentration); the Morrison Institute for Population and Resource Studies at Stanford University (the institutional home of the Diversity Project in North America); the San Francisco office of the International Indian Treaty Council; the Oakland office of the Abya Yala Fund for Indigenous Self-Development in South and Meso

America; the offices of archaeologists and population geneticists at Oxford University and Cambridge University in England; the Tuskegee University National Center for Bioethics in Research and Health Care in Tuskegee, Alabama; the National Human Genome Research Institute; and the National Science Foundation in Bethesda, Maryland. At each site, I asked for a tour and talked to people about the goals and activities of their organization or group. When visiting laboratories, I observed the conduct of research and learned about the technologies used to store human biomaterials and analyze human genetic diversity. At each site I took field notes. These notes are cited in the text using the place and date of the research.

Additionally, I attended a variety of meetings where the Diversity Project was discussed and debated. They ranged from the First International Conference on DNA Sampling (Montreal 1996) to a Special Session of the American Association for the Advancement of Science (Philadelphia 1998) to the First North American Conference on Genetic Research and Native Peoples (Polson, Montana, 1998) to the First Community Consultation on the Responsible Collection and Use of Samples for Genetic Research (Bethesda, Maryland 2000). These events gave me a rich sense of the controversy, debates, and issues surrounding the Diversity Project, as well as data on how the issues were contested, how a shared discourse at times formed, and what issues, definitions, and concepts developed as efforts to organize the Project proceeded.

Semi-Structured Interviews

From 1996–2000, I also conducted semi-structured interviews with participants in the Diversity Project debates. The interview subjects were selected to ensure that the following categories were represented:

- organizers of Diversity Project workshops and authors of central Project texts
- individuals who had been asked to represent groups Diversity Project organizers hoped to sample
- members of funding agencies asked to review Diversity Project requests for support
- members of nongovernmental organizations, and other activists, who had been commenting on and/or participating in debates about the Diversity Project
- members of governmental advisory panels who reviewed the Diversity Project
- scientists who study human genetic diversity, but were not part of the Diversity Project.

All interviews (approximately fifty in total) were semi-structured and conversational in form. Whenever possible, initial interviews were conducted in person. Many interviews generated new questions that created the need for follow-up interviews. Most of the time I conducted these follow-up interviews over the phone. When consent was obtained, interviews were taped.[267] Where possible, these interviews were taped and transcribed into machine-readable form.

Informed consent, an important consideration in any study, was a central issue in this study, because it was an object I studied as well as a practice I was ethically and legally required to follow.[268] The following informed-consent practice was followed. Each potential participant received an initial letter of invitation to take part in the study; the letter included an informed-consent form explaining the goals of my research. I used two different kinds of informed-consent forms. The first one offered participants five choices about the use of the interviews. I used this form for the first eight interviews. In the rest (and vast majority of the interviews), I used a form that explained my research and promised participants anonymity. If later in the research and writing I wanted to identify a quote, I went through a re-consent process. Where interviewees waived anonymity, I chose to use their full names only if the interviewee appears as author of a text that relates to a quotation. In the text of the book, I cite the interviews using an alphabetical and numerical code, unless using such a code would reveal the identity of an informant (see Appendix C). In the latter case I simply use the citation style (Interview [year]). All interviews were conducted by the author unless otherwise noted.

Documentary History of the Diversity Project

Two trips to the Morrison Institute for Population and Resource Studies at Stanford University, the institutional home of the Diversity Project in North America, provided me with copies of most of the Project's major internal documents. I gathered additional documents as I interviewed various participants in the debates.

Secondary Literature and Historical Sources

Prior to beginning my research, and as the research continued, I reviewed the secondary literature on the history of genetics, race, and science, and biotechnology. This body of scholarship enabled me to identify key debates. The Cold Spring Harbor Symposium held in 1950 and the UNESCO Statements on Race proved particularly important for understanding central scientific issues in debates about race and genetics as they emerged after World War II. The debates between physical anthropologists and population geneticists in the pages of *Current Anthropology* in the early 1960s helped to provide an understanding of how those debates developed and changed.

Internet Resources

As one might expect in the 1990s, much debate between organizers of the Diversity Project and members of indigenous groups took place on the World Wide

Web. For this reason, news nets, in particular Native-L, provided an important source of indigenous views about the Diversity Project.

A Final Note on Methodology

A project that consistently focuses on the question of "Who speaks?" inevitably points to the reflexive question: For whom do *I* speak? As I note at times in the text, this question came up as I conducted ethnographic research and conducted interviews. Some asked me to explain who I was, and what authorized my analysis of the Diversity Project debates. In short, questions about representation, voice, and authority were not just questions I asked others to answer, but questions I grappled with myself. My goal was not to resolve them, but rather to provide spaces in which they could be asked and reflected upon in a deliberate manner.

Appendix B
Code for Interviews

Interview A, population geneticist, 12 February 1999.

Interview B, early Project organizer, 2 July 1996.

Interview C, early Project organizer, 25 August 1998.

Interview D, Project organizer, 13 January 1999.

Interview E, early Project organizer, 2 July 1996.

Interview F, Native attorney, phone interview, 5 November 1998.

Interview G, Native scientist, phone interview, 3 April 2001.

Interview H, lawyer and bioethicist, 1 April 1999.

Interview I, Project organizer, 7 April 1999.

Interview J, advocate for indigenous rights and attendee at first MacArthur meeting, 16 April 1999.

Interview K, advocate for indigenous rights, 13 April 1999.

Interview L, biological anthropologist, 7 June 1999.

Interview M, advocate for indigenous rights, 13 June 2000.

Interview N, early Project organizer, 6 August 1999 (phone).

Interview O, advocate for indigenous rights, 19 April 1999.

Interview P, Project organizer, 20 April 1999.

Interview Q, NIH program director, 8 December 1999 (phone).

Interview R, early Project organizer, 7 April 1999.

Interview S, NIH program director, 12 January 1999.

Interview T, biological anthropologist, 20 October 1999.

Interview V, Project organizer, 22 June 1999 (phone).

Interview W, biological anthropologist, 9 September 1999 (e-mail).

Interview X, bioethicist, 12 November 1998.

Interview Y, population geneticist, 24 April 1996.

Interivew Z, early Project organizer, 13 June 1996.

Interview A1, Native attorney in health policy, 24 January 2001.

Interview A2, early Project organizer, 17 July 1996.

Interview A3, advocate for indigenous rights and attendee at first MacArthur meeting, 14 June 2000.

Interview A4, early Project organizer, 19 April 1999.

Interview A5, NIH program director, 28 July 1999.

Appendix C
Human Genome Diversity Project Time Line

Summer 1991

Diversity Project first proposed in the journal *Genomics*.

Summer 1992–Spring 1993

Planning Meetings

Population genetics, Stanford University, July 1992
Anthropology, Penn State University, October 1992
Technology and ethics, NIH Bethesda campus, February 1993

RAFI writes a letter to Diversity Project organizers expressing concerns about the Project's political and ethical implications, April 1993.

Biological anthropologist Alan Swedlund's accusation that the Diversity Project represented "21st-century technology applied to 19th-century biology" appears in the New Scientist, April 1993.

Summer 1993

Third World Network posts a message on Native-L calling for a campaign against the Diversity Project. Prompts response from Diversity Project organizers Luca Cavalli-Sforza and Henry T. Greely. Further critical messages ensue.

Indigenous groups from fourteen UN member states draft a declaration calling for "an immediate halt to the Human Genome Diversity Project."

Fall 1993

International planning meeting held in Sardinia, Italy.

Formation of regional and international planning committees, including the North American Regional Committee.

Spring 1994

The MacArthur Foundation sponsors a meeting in San Francisco designed to facilitate dialogue between Diversity Project organizers and those asked to represent Native Americans groups.

Fall 1994

Formation of the NAmC Ethics Subcommittee.

Spring 1995

Drafting of the NamC Model Ethical Protocol.

Fall 1995

Draft of the NAmC Model Ethical Protocol finished and reviewed by the Human Genome Organization Ethical, Legal and Social Implications committee.

Spring 1996

National Science Foundation Request For Proposals (RFP) for Human Genome Diversity Project Pilot Studies issued.

Fall 1996

Grants for NSF Pilot Studies awarded.

First International Conference on DNA Sampling held in Montreal, Canada.

Spring 1997

National Human Genome Research Institute holds first planning meeting to discuss genetic variation research.

Fall 1997

National Research Council committee commissioned to review the Diversity Project issues report. *Science* and *Nature* publish conflicting articles on the findings of the committee.

Winter 1998

Meeting held at Stanford University that brings together Diversity Project organizers with twenty individuals asked to represent Native American groups. Meeting reportedly ends in an impasse.

Fall 1998

A North American Conference on Genetic Research and Native Peoples held in Montana.

Those asked to represent U.S. racial and ethnic communities invited to the NIH to discuss human genetic variation research with heads of the Institutes.

1999–Present

Human genetic variation research rises to the top of agendas at several Institutes at the NIH. Debates about the meaning and proper use of racial categories increasingly trouble organizers of this research.

In April 2002, under the rubric of the Human Genome Diversity Project, the Centre pour l'Etude du Polymorphism Humaine makes available 1,064 cell lines from fifty-one populations that it claims represent most of the world's human genetic variation.

In April 2004, Cavalli-Sforza writes a letter to *Nature* that cites the CEPH collection as evidence that the Diversity Project is an active and successful project. Calls for continued financial and political support of the Project.

Notes

Chapter 1 Introduction

1. This group of scientists included Luca Cavalli-Sforza (Stanford population geneticist), Allan C. Wilson (UC Berkeley biochemist and molecular evolutionary biologist), Mary-Claire King (then a UC Berkeley medical and population geneticist), Charles Cantor (principal scientist of the Department of Energy's Human Genome Project), and Robert Cook-Deegan (medical geneticist and policy adviser for James Watson, then director of the Human Genome Project). Their call for the survey appeared in the journal *Genomics*.

2. The term "Human Genome Diversity Project" is highly contested and filled with logics and politics that could and should be explored. In this book, I use the term to refer to the definitions provided by the mostly American organizers of the Project who tried to generate U.S. support for a global survey of human genetic diversity. The Diversity Project has since been multiply defined in regions across the globe. Those who resist the Project provide some of these definitions. Others are provided by the Project's regional committees, which now exist in North America, Europe, Africa, South America, and China. Some would argue that the Diversity Project is moving forward in some of these other regions of the globe (Greely 2001).

3. Additionally, the Diversity Project has been the subject of inconclusive national and international reviews. In October 1995, the United Nations Educational Scientific and Cultural Organization's International Bioethics Committee reviewed the Human Genome Diversity Project. *Nature* magazine reported an unfavorable review. See (Butler 1995). In interviews with the author, Diversity Project leaders contested this interpretation. In October 1997, the National Research Council released its long anticipated review of the Diversity Project, *Evaluating Human Genome Diversity.* Again, an inconclusive interpretation of the re-

view resulted. *Science* reported support for the Project (Pennisi 1997). *Nature* reported a negative review. See ("Diversity Project" 1997).

4. In the mid- and late 1990s, ethicists, policy makers, and scientists involved in efforts to create projects to study human genetic variation at the National Institutes of Health attempted to distance their studies from the Diversity Project. In the last few years, however, those attempting to accommodate these new projects (such as the Environmental Genome Project) have increasingly recognized that to move forward, their projects must address similar problems as those raised by the Diversity Project (Sharp 2002, Interview S).

5. See chapter 3 for more details on King's work with the Abuelas de Plaza de Mayo.

6. In the conclusion, I will address the similarities and differences between debates stirred by the Diversity Project and debates stirred by current efforts to study human genetic variation.

7. For examples of works that exemplify this framework, see Shapin and Schaffer 1985, Ezrahi 1990, Jasanoff, et al. 1995, Wynne 1996, Hilgartner 2000, Daston 2000, Jasanoff 2004. This book has also greatly benefited from provocative discussions in the Science and Technology Studies seminar led by Professor Sheila Jasanoff at the Harvard University Kennedy School of Government in the fall of 1998 and spring of 1999. This seminar provided a space for scholars from political science, and science, and technology studies to gather to develop the emerging analytic framework of co-production that I draw upon and develop in this book.

8. On most accounts the term "ideology" itself did not emerge until 1796 when de Tracy used it to talk about a science of ideas (Williams 1983, 154). Since its emergence, it has been defined in a variety of different ways. In its most general sense in political theory, ideology is a central concept used to describe the particular ways in which discourse relates to what is political. In the Enlightenment tradition, the prevailing conception of ideology is the false-consciousness conception. In addition to holding that ideology is discourse that conceals its political or social character, this view holds that ideology functions to perpetuate dominant social interests by propagating false views about the world, or epistemic fallacies (Zizek 1994, 10). In recent years, several social and political theorists have critiqued these latter two criteria for ideology (the functionality criterion and the epistemic criterion). For further explanation of the functionality criterion and the epistemic criterion, see Sturr 1998, 17. For an overview of critiques of these criteria, see Eagleton 1991; Zizek 1994. For my work, the most important critique is the scientistic critique. This critique has been made by theorists such as Karl Mannheim who have noted the limits of a view that places ideology on one side of an epistemological divide and science on the other, implying that there is a position of critique that is free of ideology. For an explanation of this view, see Eagleton 1991, 108. This limitation, among others, has been enough for many to abandon the concept of ideology. I join a handful of other scholars in arguing that this abandonment is not necessary, and would be a loss. As the political theorist Christopher Sturr has clearly and carefully argued, we can retain a critical theory of ideology without accepting the functionality or epistemic criterion. The crucial criterion of this critical theory is the covertness criterion: the criterion that ideology is a discourse whose political character is somehow covert. The goal of critique, then, is

to make the covert (or latent) political character of discourse transparent. This study operates in this tradition of critique.

9. "Discourse" here is defined as a set of rules and practices that differentiate objects, *and* gain material institutional support. I also hold "discourse" to define the field of possibilities—what can and cannot be contemplated. This definition is Foucauldian in inspiration. For a full explanation of this meaning of the term as I use it, see Foucault 1972, 79–105.

10. Many have characterized this view of ideology as "functional false consciousness." For an overview, see Sturr 1998, 2–10. However, Frederick Engels, who coined the term "false consciousness," did not maintain that an epistemic critique necessarily followed from a false consciousness critique. For Engels, false consciousness does not result from holding false views, but from not knowing the social basis of one's views (Engels 1937 [1893], 511). I want to thank Christopher Sturr for pointing this out to me.

11. The same would be true of any category asked to take the place of "race." To gain institutional support, any category used to classify human beings for the purposes of scientific research would have to be socially meaningful. If socially meaningful, then the category will not only act to discern truth, it will have ordering effects in society.

12. For a detailed account of the emergence of the Human Genome Project, see Cook-Deegan 1994.

13. These Statements will be discussed in detail in chapter 2. For now it is only important to note that the diversity of views about race would be manifest even in the need to publish two Statements on Race.

14. These debates will be discussed in detail in chapter 2.

15. One prominent site where these different visions of what biologization (in particular, geneticization) will mean for humanity (its identity and potential) stands in great public tension today is Iceland. There, a private company, deCODE genetics, proposed to create a national database that would contain medical, genealogical, and genetic details about all the citizens of this small island nation. This proposal has stirred the passions and ire of many. Celebrated scientists such as Richard Lewontin as well as ethicists and legal scholars in the United States have joined the consumer group Mannvernd in arguing that the database would violate human rights (Coghian 1998). Others argue that the critique is unwarranted. In particular, Rabinow has argued that the moralizing tenor of critics has prevented a critical assessment of what is going on in Iceland. Critics have, he argues, fallen into the trap of humanism described by Michel Foucault: providing solutions to problems that have not been adequately posed (Rabinow 1999, 15).

16. See also the special issue of the *Social Studies of Science* (Fall 1998) on "Contested Identities."

17. Rabinow has coined the term "biosociality" to refer to this new locus of identity (Rabinow 1996).

18. Just as those argued that definitions of race based on culture would not necessarily end racism, so too these critics argue that the new purportedly liberatory categories of population genetics could just as easily be used for racist ends (Smith 1994).

19. For key legal decisions see *Cherokee Nation v. Georgia*, 1831; *Worcester v. Georgia*, 1832.

Chapter 2 Post–World War II Expert Discourses on Race

20. Stocking linked this shift to the rise of the theory of polygenism. Rather than argue that all humans came from the same monogenetic root, and then developed differences through historical and environmental processes, the theory of polygenism held that human differences derived from different physical origins (Stocking 1982 [1968], 38–39). On this polygenetic account, humans did not come from a common stock, but from different (i.e., racial) origins.

21. Thus, her history begins in 1800 and ends in 1960—an era she defines as "the modern period," a period in which "people were preoccupied by race" (Stepan 1982, x).

22. For Stepan's views on ideology, see ibid., pp. xv, 174.

23. See, for example, ibid. 1992 25, 298.

24. See, for example, the first chapter of Barkan's book for an account of the role that nationalist ideologies played in the rise of racial categories (ibid. 1992, 15–20).

25. VanHorne cites the *Milwaukee Journal* and not the UNESCO Statement. The UNESCO Statement made no such claim.

26. I use the term "Negro" here to be faithful to the terms used at the time. To change this label to one accepted by contemporary conventions would be to imply that there is a stable referent, race, for which just terminology changes. In this study I treat each racial label as an object of study that needs to be historicized and understood as the product of intertwined social and scientific ideas and practices.

27. The Myrdal study has now become the source of a large secondary literature in sociology and American history. For an exemplary analysis of the place of the study in U.S. history, and changing attitudes toward it, see Southern 1971.

28. As further evidence that U.S. government officials considered the country's racial profile a national security issue, Southern cites the State Department's request for a copy of the Myrdal study (Southern 1971, 6).

29. As the historian of the Civil Rights movement Taylor Branch has documented, this discourse of colorblindness arose at the time that the U.S. government attempted to contain and suppress the Civil Rights movement. In 1962, Branch reported, the Federal Bureau of Investigation (FBI) went so far as to assign the celebrated Civil Rights leader Martin Luther King "full enemy status" (Branch 1988, 692).

30. This effort had been assigned to UNESCO by the United Nations Economic and Social Council.

31. Many observers of the history of race and science view Montagu as one of the most committed anti-racist scientists who did the most to undermine concepts of race in science. See, for example, Barkan 1992, 339–40. However, as we will see later, it is not clear that Montagu opposed all concepts of race.

32. The broader contexts in which science would have been viewed as an appropriate response to racial prejudice will be addressed below.

33. For a historical look at British and U.S. geneticists' attitudes toward race-crossing, see Provine 1973.

34. For exemplary histories of Nazi theories of race, see Proctor 1988 and Weindling 1989.

35. They cited one of the first renowned population geneticists, Leslie Clarence Dunn, who argued that geographic isolation was the "the great race-maker" (UNESCO 1952b, 33).

36. As commentary surrounding the Second Statement would make clear, what was meant by pure race differed. For some, such as Dobzhansky, a pure race had to be genetically uniform. This kind of race, he argued, only existed in asexual species (UNESCO 1952a, 80). Others defined pure races as groups of individuals that shared as little as one gene or physical trait that did not overlap with any other group. Such a trait, if it existed, could be used to classify human populations into systematic categories (UNESCO 1952a). Later some geneticists would defend the concept of pure race as a useful construct for making sense of Mendelian experiments, but following World War II most identified it with the most pernicious of racial theories and rejected it on the grounds that races were not deterministic, but statistical entities.

37. The importance of this shift from visual to statistical inspection will become clear later in this chapter as the question "Who is an expert?" gains salience.

38. For a detailed historical and philosophical critique of the argument that race replaced population as an object of study in science, see Gannet 2001. The meaning and significance of the widely heralded rise of population genetics in biology has been explored by Haraway 1989, 199–200 and Provine and Mayr 1982.

39. Of course, in principle, there is no reason why one could not create a hierarchy based on statistical differences. Indeed, Richard Herrnstein and Charles Murray did extactly that in *The Bell Curve* (Herrnstein and Murray 1994).

40. This message begins "*What is Race?*" Specifically, Part I of this pamphlet explains how humans can both share a common ancestor and be divided into separate races. See in particular UNESCO 1952b, 11–36.

41. Muller was a member of Thomas Hunt Morgan's famous *Drosophila* laboratory where the first gene mapping efforts took place. Muller was also famous for his studies of mutagenesis.

42. For a summary of the debate over mental and emotional traits, see the University of Columbia geneticist Leslie Clarence Dunn's commentary (UNESCO 1952a, 90). For further opinions on this subject, see UNESCO 1952a, 17–35.

43. For the argument that ideas of race in science resulted from the "sensitivity of the human sciences to ideological and political factors," see Stepan 2000, 184.

44. See also Stepan 1982, ix–x.

45. For a review of the typological-population distinction as understood by historians and philosophers of biology, see the work of the philosopher and historian of biology Lisa Gannett (Gannett 2001). I thank Gannett for calling my attention to these debates.

46. For a further discussion of population geneticists' definition of race as populations, see Gannett 2001.

47. Coon's book became the focus of an already existing debate about the public authority of anthropology and genetics, and the role the discipline should play in broader debates about race in society. For a discussion of these debates, see Jackson 2000. Coon was a leading figure in physical anthropology. For a discussion of the place of his views in the history of race and science, see Goodman and Hammonds 2000, Marks 2000, and Silverman 2000.

48. As the historian and philosopher of biology Lisa Gannett has pointed out, even on this point, there are no sharp lines. Earlier in his career, Dobzhansky did entertain notions that populations could be usefully thought of as types (Gannett 1999).

49. Dobzhansky first made this distinction between a typological and population approach to the study of race in a paper he presented in the concluding session of the 1950 Cold Spring Harbor Symposium where he contrasted the concept of race as a type to the concept of race as a population (Gannett 1999).

50. Lewontin himself conceived his article in the midst of his anti-racist political work with the Black Panthers in Chicago. However, in correspondence with the author, Lewontin explained that he probably would have written the piece anyway, given that its "technical side" (the use of the Shannon Information Measure, a method for measuring species diversity developed in the field of population ecology) would have interested him even if he had not been doing any political work (Correspondence with author 25 May 1999).

51. Montagu distinguished between two different concepts of race. The first concept of race he colorfully referred to as the "omelette conception of race," which he attributed to "older anthropologists." This concept conceived of race-making as "knocking the individuals together, giving them a good stirring, and then serving the resulting omelette as a 'race' " (Montagu 1950, 318). Montagu distinguished this conception of race (as the averaging of individuals) from the population concept of race, which conceived of races as a group of populations that differed in the frequency of some gene or genes. Montagu rejected others' characterizations of his views as amounting to denying the existence of human races (Montagu 1950, 316, 318).

52. Dunn goes on to note: "Of course it may serve a taxonomic function also in enabling us to sort out and put in some order the variety we find in any collection of organisms" (Dunn 1959, 91).

53. For example, Dunn and his colleagues studied blood group frequencies of a "small Jewish community in the old ghetto district of Rome" (Dunn 1959, 110). The Institute only lasted a few years before it closed. It is unclear why it was not successful. One population geneticist who knew the scientists involved cited bureaucratic ineptitudes (Interview B). However, very few records on the Institute exist, and so the possibility of further inquiries into the reasons for the Institute's closing are limited. For a description of research proposed by those at the Institute, see Dunn 1959.

54. Dobzhansky also did not believe that just because the dividing lines between races were arbitrary that races were not real. As he wrote in a book with Dunn:

> One should not conclude, however, that because the dividing lines between races are frequently arbitrary, races are imaginary entities. By looking at a sub-

> urban landscape one cannot always be sure where the city begins and the country ends, but it does not follow from this that the city only exists in the imagination. Races exist regardless of whether we can easily define them or not. (Dunn and Dobzhansky 1946, 126)

55. For a more thorough description of Dobzhansky's views on classification, see Gannett 2001.

56. Hooton led the anthropology department at Harvard University during the 1920s when Boas and his students were founding the field of cultural anthropology. According to historians of science, Hooton focused on the study of race while Boas replaced a study of race with a study of culture. For a review of these histories, see Jackson 2001.

57. In order to move into the "present," he urged students of human variation to shift their focus from studies of phenotype to studies of phenotype *and* genotype. The only phenotypic traits that should be used in studies of race formation, he suggested, should be those traits whose underlying genetics had been worked out. At the time, blood-group was the most popular and well-known example of such a trait. William Boyd's *Genetics and the Races of Man*, a study based on blood group analysis, was published in the same year as the Cold Spring Harbor Symposium (Boyd 1950). For analysis of the role that blood group studies played in racial analysis, see the work of Rachel Silverman (Silverman 2000).

58. Population geneticists like Dobzhansky recognized that "the genic basis of relatively few human traits is known," but still argued that studying the distribution of these few traits could tell us more about the formation of human races than "a great abundance of measurements" (Dobzhansky quoted by Montagu 1950, 318).

59. Historians have labeled this era in the history of biology the evolutionary synthesis (Provine and Mayr 1982). For a definition and discussion of the synthesis, see chapter 3.

60. As Wiercinski noted: "[C]urrent typological thinking is not necessarily connected with a pregenetic phase in the development of anthropology" (Wiercinski 1964, 319).

61. For another excellent review of these debates, see the papers on controversies sparked by the publication of Carleton Coon's *The Origin of Races* (Marks 2000).

62. My work in this chapter builds upon the work of the historian of biology William Provine. In his essay, "Genetics and Race," Provine argues that belief in race differences persisted much longer than most accounts admit, and certainly did not dissolve with the publication of the UNESCO Statements. Even today, he points out, there exists only agnosticism, and not certainty, among geneticists on the subject of hereditary mental differences between races (Provine 1986).

Chapter 3 In the Legacy of Darwin

63. Rabinow and Palsson's use of this term is inspired by Yves Dezalay and Bryant Garth's analysis of the ways in which the symbolic capital generated by the "humanitarian" can be traded for economic capital (Dezalay and Garth 1996, 8).

64. After Allan Wilson died of leukemia in August 1991, many would come to view Cavalli-Sforza as the Project's intellectual leader.

65. For a detailed account of the emergence of the Human Genome Project, see Cook-Deegan 1994.

66. Not all scientists were as convinced of the merit of the Human Genome Project as Watson and Cook-Deegan. For early debates about the value of the Genome Project, see Cook-Deegan 1994.

67. For an example of the usage of the term "vanishing," see Cavalli-Sforza and Bowcock 1991, 495.

68. At the time, the United States favored Iraq.

69. See details at http://www.phrusa.org/research/chemiraqgas1.html.

70. See Cavalli-Sforza and Bodmer's *The Genetics of Human Populations* for the classic explanation of this population-based approach (Bodmer and Cavalli-Sforza 1971 [1999]).

71. One must keep in mind the broader issues about the allocation of funding in biology that were at stake at the time. Many biologists were worried that the Genome Project would lead to a massive investment in the development of new technologies that would divert funds from scientifically interesting research.

72. King conducted this research while a graduate student of Allan Wilson, a biochemist who many considered the other scientific luminary whose support would be needed to launch the Diversity Project. Wilson died of leukemia in August 1991. Mark Stoneking, Anna DiRienzo, and Svante Paabo, all colleagues of Wilson's, would continue their involvement in efforts to organize the initiative.

73. In May 1987, the Argentinean Congress passed a law establishing a National Genetic Data Bank to "solve any type of conflict that involved issues of affiliation, including cases of disappeared children" (Arditti 1999, 72). Testing was to be provided free to relatives of the disappeared, and failure to comply with the law would be considered a sign of complicity with the kidnappings. It is expected that the bank will be used to solve cases of kidnapped children until the year 2050 (ibid., 72–73).

74. Population genetics has also been attached to other liberatory causes, such as proving the innocence of prisoners on death row (Gootman 2003).

75. Cook-Deegan and King knew each other from their activities in Amnesty International.

76. Thomas A. Bass uses this phrase to describe Mary-Claire King in his chapter on her in his *Reinventing the Future: Conversations with the World's Leading Scientists* (Bass 1994, 219).

77. For an account of Nancy Wexler's work, see Alice Wexler's *Mapping Fate* (Wexler 1995).

78. Cook-Deegan included among potential relief organizations the International Rescue Committee, the International Commission of the Red Cross, Oxfam, the United Nations High Commission on Refugees, the World Health Organization, UNESCO, and Catholic and Protestant relief organizations (Cook-Deegan 1991).

79. *Genomics* was a journal established specifically to serve the human genomics community. It is not clear why the decision was made to announce the idea

in an article in *Genomics* rather than writing a grant proposal. Cook-Deegan recalled that Cavalli-Sforza did not think that the NIH would support the proposed $20 million project without prior public support in the genomics community. Publishing an article also followed the model of the Human Genome Project (Personal communication with author, August 1999). The idea for the Genome Project was first proposed by Renato Dulbecco (president of the Salk Institute) in an editorial in *Science* magazine (Dulbecco 1986).

80. In addition, a reference to the Hibakusha report remained. At the time, Cook-Deegan was reading about the social history of genetics in Japan and learning about how the geneticist Herman J. Muller's predictions of permanent DNA damage to bomb survivors had led to severe stigma among the Hibakusha, especially among the young women.

81. Evidence of this concern includes: the 1988 Office of Technology and Assessment "feasibility study" mentioned it as an issue to be discussed; two senior African Americans were asked to sit on the original working group of the Ethical, Legal and Social Implications Program of the National Center for Human Genome Research; "new eugenics" was cited as one of two examples of an "ethical issue" in the NCHGR's first five year plan; in 1989–90 the historian of eugenics Daniel Kevles (whose book many human geneticists had read) hosted a seminar series at the California Institute of Technology on human genome research. Kevles's contribution, a lecture/essay called "Out of Eugenics: The Historical Politics of the Human Genome," became the first article in that seminar's book, *The Code of Codes: Scientific and Social Issues in the Human Genome Project* (Kevles and Hood 1992). Additionally, in June 1990, James Watson received a memorandum from Benno Muller-Hill, a noted molecular geneticist and a champion of concerns about eugenics in the genome community. Attached to the memo was a paper called "Bioscience under Totalitarian Regimes," and a draft of a statement on "Lessons from Eugenics and Racial Hygiene." Muller-Hill hoped that Watson would consider both in preparation for the planned drafting of a "Declaration" of genome ethics at the second international genome meeting to be held in November in Valencia, Spain.

82. The race and IQ debate re-emerged in 1969 when Arthur Jensen, a professor of education at Berkeley, published an article in the *Harvard Educational Review* that argued that the average IQ of "blacks" was much lower than that of "whites" (Jensen 1969).

83. These concerns grew even greater in 1990 when a critical article profiling James Watson and the Human Genome Project appeared in the weekly news magazine the *New Republic* (Wright 1990). Not surprisingly the article troubled Watson and other leaders at the Center. Watson responded by calling upon the director of the newly formed ethics program at the Genome Center to make ethical, legal, and social implications (ELSI) more visible. Watson had already made his concerns about ELSI issues public and visible through announcing that 3 percent of the Genome Project's funds would be devoted to addressing ELSI issues, funds that led to the creation of the new program. But now he called upon the director of the program to initiate the planning of conferences that would make the program's efforts more visible than receiving, reviewing and awarding grants—the program's activities to this point (Interview A5).

84. Cavalli-Sforza also would be a logical choice to respond to these concerns, as on many occasions he had debated William Shockley, a Berkeley physicist who in the 1970s argued that "black women" should be given $5,000 if they agreed to be sterilized (Cavalli-Sforza 1995). Cook-Deegan would later ask Cavalli-Sforza to present his ideas about the meaninglessness of race to the first international bioethics meeting supported by the Human Genome Project. Cavalli-Sforza could not go, and a colleague, Marcello Siniscalco, spoke.

85. For a further elaboration of this view see Cavalli-Sforza, et al. 1994, 19.

86. For an elaboration of this closer-to-the-genes, closer-to-reality argument, see Cavalli-Sforza and Bodmer 1999 [1971] and my discussion later in this chapter.

87. For a description of these new technologies, see Cantor and Smith 1999.

88. However, as discussed in chapter 2, Carleton Coon and other physical anthropologists argued in the 1950s and 1960s that these visible physical traits could not be abandoned. The role of visible physical traits would also be a source of controversy in the Diversity Project case.

89. For the full text of the decision, see *Brown v. Board of Education*, 347 U.S. 483 (1954). For a detailed historical account of this pivotal case, see Kluger 1975.

90. The theorists of American society, Michael Omi and Howard Winant argued in their book *Racial Formation in the United States* that these claims that race might no longer matter in U.S. society were "outlandish" and hid from view the multiple and consequential ways in which the United States remained a color-conscious society. In this study, I analyze a parallel phenomenon in science whereby the claim that race is meaningless in society obscures continued scientific efforts to study race.

91. For a discussion of the arguments of these culturalists, see chapter 2.

92. For a review of this research, see Lewin 1982, 218–36.

93. This approach was opposed to the "hypothetico-deductive approach" which experimentalists viewed as pure speculation (Mayr and Provine 1980).

94. The particulate theory opposed the blending theory of evolution. The latter theory, the most widely held theory in Darwin's time, held that the heritable traits of parents mixed in the offspring. The particulate theory held that the inherited material was discrete physical particles that could not blend.

95. Some observers of the history of biology have argued that an envy of physics accounts for this belief that only physical forces, and not natural variation, could explain evolutionary change. At the time, physics was deemed the most prestigious and rigorous of the sciences (Keller 1992, Mayr 1980). Indeed, Morgan did write in 1932 that biology would only progress by "an appeal to the experiment . . . the application of the same kind of procedure that has long been recognized in the physical sciences as the most dependable one in formulating an interpretation of the outer world" (Morgan quoted in Mayr 1980, 10).

96. Early mathematical-population genetics, Mayr argued, tended to view evolution as simply a change in gene frequencies over time, a formulation he labeled "bean-bag" population genetics (Mayr 1980, 12).

97. Recall that Montagu is most well known for his purported rejection of the term race in favor of ethnic group. However, a careful reading of his writings reveals that Montagu made the far more specific claim that the term race should

not be used to describe biological differences in popular usage because it was prone to too many misunderstandings and conflations with what he understood to be social definitions of race. However, among his scientific colleagues, he freely asserted: "In the biological sense there do, of course, exist races of mankind" (1950, 318).

98. Mathematical genetics began when the British mathematician G. H. Hardy and the German physician Wilhelm Weinberg independently came up with a mathematical formulation for determining the frequency with which different genotypes occur in populations in which individuals breed at random (Kevles 1985, 195).

99. For many historians of biology, this replacement largely enabled the so-called evolutionary synthesis—a unification of biology forged by bringing Darwin's theory of natural selection in harmony with the principles of genetics (Provine and Mayr 1982, 1).

100. The meetings' attendees included the leading figures in both genetics and physical anthropology, including Theodosius Dobzhansky, Ernst Mayr, Ashley Montagu, George Gaylord Simpson, James Neel, Leslie Clarence Dunn, Curt Stern, William C. Boyd, and Carleton Coon.

101. The position on race that Demerec presents here is the same as the one that would be put forward in the First UNESCO Statement on Race published in the same year as the Symposium: race differences are not absolute and static, as suggested by the taxonomist's typologies, but rather dymamic, as revealed by the biologist's study of human populations.

102. This ecumenical approach broke down in practice as anthropologists critiqued geneticists for being too focused on quantitative traits, and geneticists noted the limits of qualitative methods. See, for example, Cold Spring Harbor Symposia on Quantitative Biology 1950, 28. As demonstrated in chapter 2, differences existed both between and among anthropologists and geneticists as to how to study and define races.

103. See the previous section of this chapter, "The Emergence of Population Genetics," for a description of how the idea that geneticists should study natural populations took hold.

104. The goal of the first session of the CSH Symposium was to elaborate the meaning and importance of this concept.

105. This book is considered a founding text of population genetics, and Dobzhansky is regarded as a founding father. For a review of the importance of Dobzhansky's work to the discipline of population genetics, see Provine, Lewontin, Moore, Wallace, eds., *Dobzhansky's Genetics of Natural Populations* (1981).

106. For a further elaboration of the populationist understanding of race, see Gannett 2001.

107. Most biologists believed that humans had remained relatively stable in their racial groups until the Age of Transportation (boats) facilitated global mobility. For further details of this argument, see Cavalli-Sforza, et al., chapter 1, 1994.

108. Cavalli-Sforza and Bodmer would later explain in their 1971 textbook, *The Genetics of Human Populations*, that the existence of human races indicated that the human species had been on its way to splitting into several different species when recent increases in communication and migration intervened and pro-

duced the current state of interracial mixing (Cavalli-Sforza and Bodmer 1971, 698).

109. Participants also called attention to many of the difficulties associated with conducting population studies, such as defining the population and a correct sampling strategy. Although most agreed that population studies should be done, much debate remained about how these studies should be done, and how they related to previous studies of race (see chapter 2 for debates among physical anthropologists and geneticists about the meaning of race and the value of studying populations). How these debates played out in the case of the Diversity Project is the subject of the next chapter.

110. Like the Diversity Project, this research would also become the source of controversy in the 1990s. See chapter 6 for details.

111. They believed that these groups would provide the kind of clear demarcations around groups that demographer's enjoyed.

112. See, for example, Cold Spring Harbor Symposia on Quantitative Biology 1950, 22–23, 159, 165–73. CSH participants were not the only ones using the term "isolate." Indeed, it had been widely used for decades in genetics.

113. Up until the 1950s, scientists only used the fossil record to answer questions about human evolution (Jones 1993, 100–101). These fossil studies roughly date back to 1856, when the first hominid fossil, a Neanderthal, was discovered. Indeed, as late as the year 2000 the text of the "The Human Fossil Record" exhibit at the American Museum of Natural History read: "Only the fossil record can tell us exactly how and when these transformations [from *H. erectus* to the Neanderthals] in humans occurred (author's fieldnotes)." The value of the fossil record in studies of human evolution has been the source of great debate. For a contrasting view of the value of the fossil record, see the works of George Gaylord Simpson, Harvard professor and curator of Fossil, Mammals, and Birds in the American Museum of Natural History during the 1940s and 1950s. In particular, see *The Meaning of Evolution* (Simpson 1950).

114. In *Genes, Peoples and Languages* published in 2000, Cavalli-Sforza supplements his account of brain growth with the qualification that it was "not enough" to account for emergence of the human species. Self-consciously using computer terminology, Cavalli-Sforza notes that in addition to this "hardware," human evolution required improvements in "software." By "software," Cavalli-Sforza means language (Cavalli-Sforza 2000, 59).

115. For a discussion of the concept of miscegenation and its role in structuring American legal concepts of race, see Pascoe 1996.

116. Here I want to distinguish the population geneticist organizers (namely Luca Cavalli-Sforza and Walter Bodmer) involved at this point in the planning of the initiative from the medical geneticist supporters (namely Robert Cook-Deegan and, to some extent, Mary-Claire King). As this chapter demonstrates, the former group wanted to study race while the latter had no particular interest in questions of race formation, although they might use conceptual orders structured by notions of race.

117. To be sure, "Negroid" did not fall out of use. It, for example, appeared in the following line in the summary of an article by Mark Stoneking (a student of Allan Wilson's) that appeared in the *American Journal of Human Genetics* in

1996: "Within Africa, the deletion was not found among Khosian peoples and was rare to absent in western and southwestern African populations, but it did occur in Pygmy and Negroid populations from central Africa and in Malawi and southern African Bantu-speakers" (Stoneking 1996). When I later asked Stoneking about the category Negroid he responded that he thought the category was scientifically meaningless and that he did not use it. He was surprised when I pointed out to him that it did appear in this article (Interview with author, Penn State University, 25 August 1998). I argue that this lack of recognition of a use of a racial category is symptomatic of the effects of ideology described by the social theorist Slavoj Zizek. According to Zizek, we enter the realm of the ideological when a claim or statement "is functional with regard to some relation of social domination ('power', 'exploitation') in an inherently non-transparent way" (Zizek 1994).

118. For a historical study of the rise in cultural power of molecular biology, see Kay 1993.

Chapter 4 Diversity Meets Anthropology

119. During the 1970s some physical anthropologists began to refer to themselves as biological anthropologists. This change in name was intended to reflect a change in approach similar to that described by the population geneticists. These anthropologists argued that the term denoted a shift to a focus on dynamic biological processes, and not static physical forms (Interview T). For simplicity, however, in this book I use the terms "physical anthropology" and "physical anthropologists" unless it is important to my argument to make a distinction between the terms "physical" and "biological."

120. More than the rule of law, Foucault believed that the rules that govern what can and cannot be discussed, what can and cannot count as knowledge, constitute the formative elements of modern regimes of power. These rules define the conditions for the possibility of a science, conditions that Foucault labeled an *episteme* (Foucault 1966, xxii). In the current episteme, Modernity, Foucault found the human at the center, a unifying element from which knowledge flowed. Thus, theories about the human prove particularly important both to the constitution of knowledge and to power in this era. For Foucault's explanation of the context in which he formulated the problematic of power-knowledge, see Foucault 1980.

121. This approach drew upon Wilson's experiences conducting the mitochondrial Eve study—a study famous for its claim that all humans trace their ancestry back to "one woman who is postulated to have lived 200,000 years ago, probably in Africa (Cann, Stoneking, and Wilson 1987). This study used mitochondrial DNA (mtDNA) collected from individuals, not nuclear DNA collected from populations. Mitochondrial Eve researchers cited two reasons for studying the mitochondria of individuals. First, mtDNA mutates ten times faster than nuclear DNA, and thus genetic differences are more plentiful and can be detected at the individual level. Second, this DNA is inherited from the mother only. Thus it avoids

the mixing problem created by the recombination of nuclear DNA, mixing that obscures the history of individuals.

122. The National Research Council later discussed the strengths and weaknesses of these approaches in its 1997 report, *Evaluating Human Genetic Diversity* (National Research Council 1997).

123. Allan Wilson died of leukemia in August 1991. Mark Stoneking, Anna DiRienzo, and Svante Paabo, all colleagues of Wilson's, attended the July 1992 Stanford meeting.

124. As explained in chapter 3, these scientists defined the population as a group of "interbreeding individuals who share a common pool of genes . . . transmitted from one generation to the next according to Mendel's laws" (Bodmer and Cavalli-Sforza 1971, 39).

125. For a description of the clinalist critique, see chapter 2. These debates continue today as proponents of the International Haplotype Map (the NIH's effort to map human genetic variation) struggle to define a sampling strategy for their initiative (Weiss and Kittles 2003).

126. By "an anthropologist's perspective" Roberts and others writing about the Project at this time meant the perspective of a physical anthropologist. The validity of this equation of anthropologist with physical anthropologist, as well as the nature of physical anthropology, would later come into question.

127. Members of the committee were Julia Bodmer (Imperial Cancer Research Fund, UK), L. Luca Cavalli-Sforza (Genetics, Stanford), Kenneth Kidd (Genetics, Yale), Mary-Claire King (Epidemiology, UC Berkeley), Svante Paabo (Genetics, University of Munich, Germany), Alberto Piazza (Human Genetics, University of Turin, Italy), Marcello Siniscalco (Imperial Cancer Research Fund, London), Jared Diamond (Physiology, UCLA School of Medicine).

128. For an articulation of this "science first" approach, see the report written by the National Research Council (NRC) committee appointed to review the Human Genome Diversity Project. This NRC committee decided that it could not review the Diversity Project because the project had not yet "sharply defined" its proposal for research. The scientific objectives, they argued, remained "elusive" (National Research Council 1997, 1).

129. The most critical formulation of this concern would be expressed by the physical anthropologist Jonathan Marks who argued: "The HGDP, as it turned out, had been conceived, designed and organized by molecular population geneticists with largely a folk knowledge of anthropology. It was the equivalent of teenagers trying to build a cyclotron in their backyard" (Marks 1995, 72).

130. Mark Weiss received his Ph.D. in anthropology from UC Berkeley in 1969. His thesis adviser was Sherwood Washburn, the champion of the "new physical" anthropology which sought to make studies of primate behavior a bridge between the biological and social sciences (Haraway 1989, 187).

131. All interviews are with the author unless otherwise noted.

132. Kenneth M. Weiss and Mark Weiss are not related. To avoid confusion, from here on out I will use their full first and last names.

133. In the NIH/NSF/DOE conference grant proposal the description of this "Anthropology" workshop began: "Many anthropological objectives can be furthered and modernized by the HGD initiative." This represented perhaps the first

time that an initiative document had acknowledged anthropology as a source of objectives and not just a part of the collecting infrastructure (Human Genome Diversity Project 1992c, 8).

134. The metaphor "on board" is used in the Anthropology Workshop report and the title of the Roberts article on this workshop: "Anthropologists Climb (Gingerly) On Board" (Human Genome Diversity Project 1992b, Roberts 1992). This formulation also conveyed a sense that the Project had already been constituted, and so anthropologists could only join, and not play a formative role, in constituting the Project.

135. The meeting was held 1–4 April 1992 at the Riviera Hotel and Casino in Las Vegas, Nevada. For more information on the meeting see the *American Journal of Physical Anthropology*, Annual Meeting Issue, Supplement 14, 1992.

136. Organizers attributed this separation to the requirements of ethics and expertise: anthropologists, and not population geneticists, had the appropriate skills and qualifications to ethically and appropriately sample populations.

137. The farmers also proved less than an optimal source of support for collecting blood samples. As Cavalli-Sforza recalled: "Farmers claimed that they were masters of the Pygmies and considered Pygmies their hereditary servants, but they actually did not have enough influence to convince them to give us blood if the Pygmies were not of a mind to do so." The problem of collecting blood samples would later be solved. As Cavalli-Sforza recounts: "We offered the Pygmies simple presents that I had learned were highly regarded by them, such as salt, soap and cigarettes, as well as medical treatment, within our means, for the sick." As we will see in chapter 6, these kinds of relationships with research subjects would later raise explicit ethical and political questions. In this 1986 volume, however, these questions do not arise. The only reference to the ethics of blood collection appears in Cavalli-Sforza's account of who drove the cars for his winter 1967 expedition to Africa: "Four of us drove the cars from Douala to Bangui: Vincenzo Pennetti, M.D., from Rome; Guglielmo Marin, now professor of biology in Padova; X.Y., a young technician of the Genetics Department in Pavia and our expert in automobile repairs (who does not want to be named, since he has adopted a new religion that forbids the collection of blood); and myself" (Cavelli-Sforza 1986, 3–4, 6).

138. Based on his own ethnographic research, Turnbull wrote a famous book about the Pygmies, *The Forest People* (Turnbull 1962).

139. Turnbull also appreciated Cavalli-Sforza's research. As he wrote in a chapter he prepared for a volume devoted to research supported by Cavalli-Sforza's research trips to Africa: "One further general observation should be added about the present state of knowledge and the degree to which generalizations about 'African Pygmies' can be made. To my mind, the only area investigated with the academic rigor that makes such generalizations valid is the area of physical anthropology, in particular the kinds of studies reported in this volume" (Turnbull 1986, 105). Turnbull critiqued his fellow social and cultural anthropologists for making generalizations about the Pygmies without conducting sufficient research (ibid., 103–105).

140. Like many of the most prominent students of human diversity—including Luca Cavalli-Sforza, Theodosius Dobzhansky, and Ashley Montagu—Neel be-

lieved that one could glean insights about human origins and evolution by studying the gene pools of indigenous groups, groups that Neel referred to as the last remnants of "precivilized man" (Neel 1994, 119). Neel's research on the Yanomami would also become controversial. See chapter 6 for details.

141. Maybury-Lewis later helped to found Cultural Survival, a nongovernmental organization created in 1972 to "[defend] the human rights and cultural autonomy of indigenous peoples and ethnic minorities." The organization devoted a whole issue of its magazine to the Diversity Project debates in 1996. For Maybury-Lewis's history of the organization, see www.cs.org.

142. The argument that anthropologists approached the study of human diversity and evolution with a broad range of analytic skills, and that geneticists were limited by their narrow focus on genes, echoed arguments made by the physical anthropologist Carleton Coon decades earlier. See chapter 2 for details.

143. The Wenner-Gren Foundation is a private nonprofit organization that supports the discipline of anthropology. Attendees at the Cabo San Lucas meeting included George Armelagos (physical anthropologist, Emory University), Michael Blakey (physical anthropologist, Howard University), Carlos E. Coimbra Jr. (medical anthropologist, National School of Public Health, Rio de Janiero, Brazil), Arturo Escobar (social-cultural anthropologist, University of Massachusetts, Amherst), Alan Goodman (biological anthropologist, Hampshire College), Fatimah Linda Collier Jackson (biological anthropologist, University of Maryland), Debra Martin (biological anthropologist, Hampshire College), William Roseberry (social-cultural anthropologist, New York University), Ricardo Ventura Santos (biological anthropologist, National School of Public Health, Rio de Janiero, Brazil), Sydel Silverman (then president of the Wenner-Gren Foundation), Alan Swedlund (biological anthropologist, University of Massachusetts, Amherst), and R. Brooke Thomas (biological anthropologist, University of Massachusetts, Amherst). Most who attended the meeting shared a particular interest in exploring the ways in which political and economic conditions shape the biological and cultural well-being of human populations.

144. At the Diversity Project's Anthropology workshop geneticists represented the biological perspective while archaeologists and anthropologists of all kinds represented the cultural perspective. The importance of these differences in who represents biological and who represents cultural expertise will soon become clear.

145. See chapter 2 for more on Stocking's argument.

146. In collaboration with Marcus Feldman, Cavalli-Sforza developed an approach to culture that modeled culture on evolution. This model postulated that cultural transmission could be understood through the use of genetic models (Feldman and Cavalli-Sforza 1981).

147. Of course, colonialism continued in other non-official forms.

148. As Kuklik notes, it was "once fashionable to argue that the theories of anthropology represent nothing more than types of imperial ideology" (Kuklick 1991, 26).

149. Concerns that governments could appropriate anthropological research emerged again during the Vietnam War when accusations arose that security forces were using anthropological research to help identify bombing targets. The resulting controversy—which became known as the Laos affair—provided the

impetus for the formation of the Ethics Committee of the American Anthropological Association (AAA). The current controversies surrounding the publication of the journalist Patrick Tierney's *Darkness in El Dorado: How Scientists and Journalists Devastated the Amazon* demonstrate that these concerns have not subsided. In this controversial book, Tierney argues that anthropologists such as Napoleon Chagnon (as well as geneticists such as James Neel) negatively affected the Yanomami through their research, disrupting their culture and even killing off members by administering vaccinations that contained deadly pathogens. An excerpt from the book appeared in *The New Yorker* in October 2000 (Tierney 2000). A few weeks before the publication of the excerpt, the past chair of the AAA's Committee for Human Rights, Leslie Sponsel, along with Terry Turner, an anthropologist at Cornell University and longtime critic of Chagnon, sent a letter to the president and president elect of the AAA. In their letter, Turner and Sponsel supported Tierney's claims and urged the leadership of the AAA to respond promptly to the "impending scandal." In the weeks that followed, the letter spread rapidly over the internet and a controversy broke out. While Turner and Sponsel supported Tierney, others challenged the validity of his claims. Tierney did withdraw the genocide claim, but maintained that Chagnon's research was detrimental to the Yanoamami (Zalewski 2000). On 9 February 2001 the AAA launched a formal inquiry into the allegations http://www.aaanet.org/press/ebmotion.htm (accessed 31 January 2002). In November 2003, the American Anthropological Association adopted a referendum that repudiated the accusation that James Neel and Napoleon Chagnon started a lethal measles epidemic by administering vaccinations. See http://www.aaanet.org/stmts/darkness_in_eldorado.htm (accessed 2 April 2004).

150. For a description of the culturalists, see chapter 2.

151. For a classic statement of the logic behind studying the biological makeup of so-called primitives or precivilized man, see Neel 1994, 118–119.

152. Most prominent among these critiques were Edward Said's *Orientalism* (Said 1979) and Talal Asad's *Anthropology and the Colonial Encounter* (Asad 1973).

153. Some physical anthropologists criticized the Diversity Project for not inviting African American physical anthropologists, such as Blakey and the biological anthropologist Fatimah Linda Collier Jackson, to the Penn State Anthropology workshop (Interview C).

154. This Wenner-Gren meeting had no relationship to the November 1992 Wenner-Gren meeting in Cabo San Lucas other than that it was funded by the same foundation. It was also part of the same effort by Sydel Silverman to contribute to efforts to integrate biological and cultural anthropology before her term ended as Wenner-Gren Foundation president (Interview T).

155. Moore had previously worked with Kenneth Weiss on a study of diabetes in American Indian populations in Oklahoma.

156. Despite this stated goal, the majority of those invited to the Wenner-Gren conference were physical/biological anthropologists.

157. Cavalli-Sforza and Swedlund did establish friendly enough relations to enable both to attend the conference. At the conference Swedlund made efforts to explain that his role as critic was not one that he had chosen, but one that had

been in many ways imposed upon him by the media. Indeed, he expressed regret that the *New Scientist* had highlighted his comment in its April issue. However, Swedlund also did not retract his statement. Instead, in the paper he presented to the conference he explained "the underlying thinking that had motivated [it]" (Swedlund 1993, 1).

158. Moore later became chair of the North American Regional Committee of the Diversity Project. In this article he draws upon both his anthropological training and his experiences growing up in the South and attending an all-white high school to ground his claims to expertise about racism. The latter experience, he claimed, helped him to understand the problem of scientific racism. If even his fellow all-white classmates experienced devastation when they learned of low IQ scores, then blacks who also suffered from scientific racism must have felt doubly devastated (Moore 1996, 218).

159. For a discussion of Montagu and his work as rapporteur for the UNESCO Statements on Race, see chapter 2.

160. Moore explained in another article that migration theory assumed that ethnic boundaries (or *ethnos*) had been maintained over the course of human history and that biology and language and culture had evolved together. Moore referred to this dominant model of human evolution and origins as the cladistic model (Moore 1994, 13). He opposed this cladistic model with his ethnogenetic theory. According to this theory, ethnic boundaries of human societies had not been maintained and passed on from one generation to the next, but rather had fragmented and shifted, even within one generation.

161. Moore's belief that scientists did not engage in evaluations of human character traits is countered by contemporary and historical evidence. Although none of the genes Cavalli-Sforza et al. considered in their compilation of data from past human population genetics studies, *The History and Geography of Human Genes*, connected to behavioral traits, in his comments at the February 1993 ethics workshop Cavalli-Sforza left the door open to the possibility that future studies would consider behavioral traits. In particular, in reference to the new data the Diversity Project would collect, he reportedly argued:

> What harm can come out of data of this new type [molecular]? A trait like IQ is a product of perhaps 200 genes, which are likely to interact in complicated ways and, in addition to react to a poorly known set of external, non-genetic factors and events. It is difficult to believe that a satisfactory analysis of such complicated situations regarding behavioral traits will be possible in the near future, perhaps not even in time for the younger people in this audience to see. (Human Genome Diversity Project 1993, 26).

Not "in time for the younger people in this audience to see," but never? Although Cavalli-Sforza admits that the genetic determination of these characteristics at present is based on "soft evidence," it is clear from his past writings that he was committed to the view that these differences do exist. Indeed a passage in the classic textbook *The Genetics of Human Populations* states: "There is undoubted evidence that socially important traits like intelligence and mental balance are to a fair extent genetically determined" (Bodmer and Cavalli-Sforza 1971, 770). Although Cavalli-Sforza and Bodmer wrote this statement in 1971, there is little or no evidence to indicate that they have changed their view that genetics in some part contributes to intelligence.

162. Moore does not provide specifics about the kind of anthropologist he deemed important.

Chapter 5 Group Consent and the Informed, Volitional Subject

163. My investigation of the construction of ethics builds upon an emergent body of scholarship that explores the professional domain of "bioethics" as a site in figuring the emergence of genomics. See, for example, Rabinow and Palsson, forthcoming; Shostak 2003. My work in this area also has been inspired by speakers in the Generating Values(s): Emergent Forms of Ethics in Envisioned Biocapitalist Landscapes at the 2001 Social Studies of Science Annual Meeting (organized by Kaushik Sunder Rajan and Chloe Silverman). The emerging importance of bioethics and bioethicists has also recently been acknowledged in the popular press (Stolberg 2001).

164. The exact causes of the concern remain unclear. Some cite a mishandled interchange between an NIH official and Jeremy Rifkin—a noted science critic, and a particular opponent of biotechnology—about the Diversity Project. Others argue that the NIH's experiences with the Human Genome Project had generated sensitivity to ethical problems raised by human genome research (Interview E).

165. This addition of a day devoted to ethics departed from the original goal of the planning workshops. These workshops, as described in the Summary Document for the Diversity Project, had been designed to "prepare the scientific case for the project" (Human Genome Organization 1993).

166. For the minutes of this meeting, see Human Genome Diversity Project 1993.

167. In addition to presenting the Project as an opportunity, and not a liability, some who attended the meeting left with the impression that Project leaders refused to consider that their initiative could be anything but beneficial. Any worries that the Project might lead to a new racism, and not combat the old one, were set aside. According to one prominent member of the ELSI community, the meeting was clearly just "window dressing" designed to give the Project a visible ethical presence (Personal communication with author, May 1999).

168. As explained in the Diversity Project's Summary Report, the aim of the Penn State Anthropology workshop was to "identify those issues most pertinent to the selection of representative populations from each area of the world and to begin to propose examples of particular populations that would meet the defined criteria" (Human Genome Organization 1993, 9). A list of over seven hundred proposed populations resulted. See Roberts 1992b, for a description of the meeting.

169. Indeed, in more than one interview, organizers of the Diversity Project confided that the controversies had prompted them to read books and talk to other scholars about the construction of social identities—books and conversations they might otherwise never have benefited from (Interview B, Interview C, Interview T). In important ways, the Diversity Project debates provided an opportunity for social learning. I will return to this point in the conclusion.

170. Others central to the events of 1977, and beyond, prefer to refer to events since the 1970s as evidence not of a movement, but of an enduring and essential indigenous consciousness (Interview M).

171. These definitions of nature are presented in *A Basic Call to Consciousness*, a report put together by Akwesasne Notes (a major publishing press of the American Indian Movement) on the 1977 UN NGO meeting. Many recognize this meeting as the official beginning of the Indigenous Rights Movement (Akwesasne Notes 1995 [1978], 48). The equation of indigenous people with nature employed in this text is not uncommon and would, at various points in the Diversity Project debates, provide a point of agreement between Diversity Project organizers and members of indigenous rights groups. However, as we will see, these equations of indigenous people with nature were embedded in different contexts that rendered them distinct from one another. Attempts to build common values and goals based on them would require much work.

172. The role of "the myth of the Vanishing Indian" in legitimating white conquest of the Americas has been widely critiqued. As Jace Weaver explains: "Extinction is a superior means of creating indigeneity. If all the indigenes are dead, there is no one to dispute the claim. In fact, guilt for wrongs done to the indigenous peoples in the past does not allow them to be other than *of* the past. Thus the myth of the 'Vanishing Indian' was born. . . . By viewing the Indian as vanishing and the Indian cultures as disintegrating, it was possible to view 20th-century Indians who refused to vanish as degraded and inauthentic and to contrast them with stereotypes of the 'pure,' 'authentic' bon sauvage or sauvage noble of the past and thus keep Indians safely in the stasis box of the 19th century" (Weaver 1997, 17–18).

173. Ironically, use of the term "Isolate of Historical Interest" represented an attempt by Diversity Project organizers to respond to political "sensitivities" generated by the use of certain terms. As the report of the Anthropology Workshop explained:

> [T]here is no fully acceptable way to refer to populations that are in danger of physical extinction or of disruption as integral genetic units (gene pools); some existing terms such as "endangered" populations can have various connotations. Many populations around the world, especially isolates living traditional lifestyles, will soon disappear as independent units, because of disease, economic or physical deprivation, genetic admixture or cultural assimilation. In this Report, we refer to such groups as "Isolates of Historic Interest" (IHI's), because they represent groups that should be sampled before they disappear as integral units so that their role in human history can be preserved. (Human Genome Diversity Project 1992b, 5)

174. As part of an effort to recognize and promote the rights of indigenous groups in the midst of the 500-year celebration of the "discovery" of America, in 1993 the United Nations announced the beginning of the Decade of Indigenous Peoples (General Assembly Resolution 48/163 of 21 December 1993).

175. Ward Churchill succinctly describes the importance of how others describe and view groups:

> [H]ow a group is seen by others, and how it sees itself, are factors that in many ways define the conditions under which the group will live, and the options it will be able to exercise in affecting these conditions. It is, for example, one thing to see oneself as being part of a social or political movement, quite another to be lumped in as a member of a mere "gang." Or, to take another illustration, there is a very different connotation to being described as a "law enforcement officer" on the one hand, and being branded as part of "a mob of thugs" on the other (albeit, in practice, its often hard to tell the difference between the conduct of the two groups). Hence, it seems self-evident that how individuals and groups are labeled or named—and, perhaps more importantly, how they name themselves—is vital to the circumstances of their existence. In naming ourselves, both individually and collectively, we in effect name not only our reality, but also our destiny. (Churchill 1994, 293)

Diversity Project organizers would on many occasions dismiss as insignifcant the language (such as "primitive") that Cavalli-Sforza in particular used to describe indigenous populations, arguing that this language represented nothing more than an artifact of Cavalli-Sforza's European upbringing. To this, Churchill would respond:

> [T]here is an umbilical connection between the description imposed upon any group and how it is treated, between the label a group can be convinced to accept as appropriate to itself and the treatment it is ultimately entitled to demand. This is neither an "abstraction" nor "past history." (ibid., 327)

176. In the fall of 1993 Greely would become the chair of the ethics subcommittee of the North American Regional Committee of the Human Genome Diversity Project.

177. The World Council of Indigenous Peoples formed in 1975 as an international organization with Non-Governmental Organization status at the United Nations. The founding goal of the WCIP was to fight for the acceptance of concepts of aboriginal rights at the international level as basic political and economic rights of indigenous peoples (Sanders 1980).

178. As we will see in the next chapter, one central issue at stake in the debates between Diversity Project organizers, such as Greely, and leaders of indigenous rights organizations, such as Cayuqueo, was who could represent the interests of indigenous people. This issue will be addressed in more depth in later parts of this chapter and in chapter 6.

179. In the event that any commercial product developed from a sample collected by the Project, organizers proposed requiring any user to provide some kind of payment (i.e., a royalty) to the Project. These royalties would then be used "in some way for the benefit of the sampled populations." Options considered by Project organizers included donating the royalties "to an existing organization (like UNESCO, for example), returning them to the host governments, or applying them directly to the sampled populations" (Greely 1993b). RAFI responded that the issue of how royalties were handled would have to "be determined by indigenous peoples organizations," and not the Diversity Project (Mooney 1993b).

180. This representative addressed the Council for twenty minutes, briefly describing what the Project wanted to do and why, as well as the problems organizers anticipated and some proposed solutions. One goal of his talk was to explain why Diversity Project organizers thought indigenous groups should not oppose the Project (Interview 1999).

181. This argument that political leaders in the South construct the North as a colonial power in order to further their own political agendas has a longer history. Consider a report about a study on North-South relations commissioned by the Council on Foreign Relations in the 1970s. According to the report's author, Roger Hansen: "As long as political leadership in these countries [in the South] require perceived external causes to explain domestic difficulties, and so long as 'threats' are needed to produce internal cohesion, the (Northern) ex-colonial or 'neocolonialist' powers—not the countries of the East—will in most cases best fill the role" (Hansen 1979, 5–6).

182. An organizational structure for the Project emerged after the Sardinia meeting, where, according to the HUGO report, "participants, who represented all the regions of the world and included many of those who would be most actively involved with the project," discussed a proposal outlining the Project's structure, and approved it "without dissent." Sir Walter Bodmer, then HUGO president, presented the proposal to the HUGO Council in November 1993. The structure was approved by the Council at its meeting in January 1994 (Human Genome Organization 1993).

183. The NAmC hoped that one hundred indigenous populations would be sampled as part of a pilot study (Interview B).

184. In 1992 several civic organizations in the United States celebrated the 500-year anniversary of the "discovery" of the Americas. For obvious reasons, indigenous groups opposed these celebrations. The MacArthur Foundation, a foundation in the United States known to fund "frontier" areas, supported their efforts. In particular, the MacArthur Foundation provided funds to support the creation of communication networks among indigenous people that would enable these groups to respond to the 500-year celebrations.

185. The addendum outline plans to hire a communications consultant who would, among other things, arrange face-to-face meetings with indigenous groups, including any who requested one.

186. Despite this change in goal, those responsible for drafting the Protocol did not alter substantially. Most of the scientists involved in the planning of the Project were members of the NAmC and would write the Model Ethical Protocol. This group included Henry T. Greely (Stanford University law professor), Kenneth K. Kidd (Yale population geneticist), Kenneth M. Weiss (Penn State molecular anthropologist), Catherine Twinn, Esq. (lawyer, Sawridge First Nation), Marcus Feldman (computational biologist, Stanford University), John Moore (social-cultural anthropologist, University of Florida), Russell Thornton (demographer/anthropologist, UCLA, Cherokee), Emoke Szathmary (then a physical anthropologist at the University of Western Ontario), and Ryk Ward (then a biological anthropologist at the University of Utah). Mark Weiss (then a physical anthropologist at Wayne State University) participated in the early drafting of the document before leaving the NAmC. Greely was the principal drafter of the document and

received assistance from Kidd, Moore, Twinn, and Kenneth Weiss. A draft of the Model Ethical Protocol would be made available in October 1995 (North American Regional Committee 1995). This draft was later published with very few changes in the *Houston Law Review* (North American Regional Committee 1997).

187. The central right of the research subject in the American context is that of informed consent. All subjects are held to have the right to make an informed choice about whether to participate in research. Much biomedical ethics in the United States focuses on how to protect this right. See, for example, Stanley and Stanley 1982, for efforts to think about how to best protect the informed consent rights of individuals in different contexts.

188. See chapter 4 for a description of this Wenner-Gren meeting.

189. The National Commission for the Protection of Human Subjects of Biomedical and Behavioral Research issued the Belmont Report in 1979. The Belmont Report outlines the major conditions of informed consent: consent must be given voluntarily by competent and informed individuals (National Commission for the Protection of Human Subjects of Biomedical and Behavioral Research 1979). *Moore v. Regents of the University of California* expands the concept of "informed" to include disclosure of researchers' knowledge of their interests.

190. It should be noted that it does not follow that individual consent was deemed irrelevant. Group consent was not in place of, but *in addition to*, individual informed consent (North American Regional Committee 1995, 8).

191. Note also the switch in language in this quote. The easy substitution of "populations-based" for "group" illustrates the slipperiness of the latter term, and the way in which it easily crosses the boundary between biology and society.

192. Even further, European organizers viewed the whole project of ethics to be a "North American idea" that had no place in Europe (Interview E). This is a major reason why the Model Ethical Protocol, including the group consent provision, was never adopted by the Project's international Executive Committee.

193. To address problems created by outsiders defining groups as well as the terms of what counts as adequate protection of those groups, many tribes are now writing their own research codes of conduct. For examples, see Akwesasne Task Force on the Environment 1999, and Indigenous Peoples Council on Biocolonialism 2000.

194. Beginning in the late 1990s, however, researchers and bioethicists working on human genetic variation research did begin to issue calls for "community review" (Sharp and Foster 2000). Community review, however raises similar problems as group consent. In particular, to date these efforts to implement community review have not considered the fundamental issues of power involved in defining the community involved in review. I will return to this problem in the conclusion.

195. Michael Sandel is a professor of government at Harvard University, and currently a member of the President's Bioethics Council. For the announcement of the Council's committee members made by President George W. Bush's press secretary, see http://www.whitehouse.gov/news/releases/2002/01/20020116-9.html (accessed 10 June 2002).

196. For an overview, see Smith 1998, 116–18.

197. For simplicity, from here on I will refer to these scientists as "African American biological anthropologists" and "African American geneticists" to denote the importance they would ascribe to the qualifier "African American" in their own work.

198. In this chapter, the qualifiers "African American," "Mexican American," "indigenous," "Native American," and "Caucasian" used to describe both objects of biological study and knowing subjects are not my own, but those involved in the debates about the Diversity Project. These qualifiers are themselves the object of my research. In this chapter I explore when, where, and why they are used, and the debates these uses spark.

199. Thus, they would not just be the objects of research (the known), but also the subjects of research (the knowers).

200. Recall that the MacArthur Foundation had made communication with Native and indigenous communities a requirement if the Project was to receive MacArthur funds. See chapter 5 for details.

201. In documenting the problems with this turn to "democracy," the chapter contributes to a new wave of critical scholarship on participation in science. While recognizing the potential benefits of making science more representative and participatory, this literature demonstrates that the merits of these efforts can only be assessed by taking into account the particular scientific, historical, and political conditions in which they are proposed (Epstein 1991, Epstein, forthcoming, Kleinman 2000, Montoya 2001). To date, most efforts to include diverse peoples in science—within both the academy and public policy circles—have not addressed these issues. Instead they have equated such efforts with a more "democratic science" (Sclove 1996, Kitcher 2000). In so doing, they have often opened more questions than they have answered about the meaning and goals of participation. For example, the philosopher of science Phillip Kitcher argues that we are in need of a "well-ordered science"—a science that accommodates democratic values. To achieve this kind of science, he proposes that "people with a variety of perspectives" be provided with "information" that will enable them to make decisions about what research should be funded. Further, he asserts that the "needs of women, children, members of minorities, and people in developing countries" must be addressed (Kitcher 2000, 117–35). These claims leave unanswered the following fundamental questions: Who are the "people with a variety of perspectives" required by a more "democratic science," and what determines the meaning and value of their participation? What are "the needs of women, children and minorities, and people in developing countries" who have been under-represented in science, and who decides? What constitutes "information" that "the public" will use to make informed decisions? These kinds of questions prove central to the Diversity Project debates.

202. Additionally, Cavalli-Sforza hoped local workers would be enrolled to work at regional collection centers established by the Project. As he stated in an address to a special meeting of UNESCO in September 1994: "Local participation in all areas of the world will be essential and the success of the project will be entirely dependent on international collaboration and cooperation" (Cavalli-

Sforza 1994). Samples needed to be transformed into cell lines within a day or two, and often transportation alone took a day or two from some of the remote places in which Project organizers proposed sampling. Thus, Cavalli-Sforza hoped that this work could be done locally.

203. The possibility that the issue was not just one of the appearance of the Project, but of its material structure, is not one organizers acknowledged at this time.

204. The proposal for an International Forum and the regional committees came out of this Sardinia meeting (Cavalli-Sforza 1994, 4).

205. Although the majority of the participants of the Sardinia meeting were from England, Italy, and the United States, individuals from Australia, Brazil, Costa Rica, Finland, France, Germany, India, Israel, Japan, Kenya, the Netherlands, Pakistan, Russia, South Africa, and Switzerland also attended. Reflective of how ideas of who should participate had changed in some ways and not in others, all of the attendees were scientists. For the history of how biomedical research emerged as an important site for struggles for racial equality in the late 1980s, see Epstein, forthcoming.

206. The qualifiers "Caucasian" and "African American" used here to describe DNA and genomes are not my own, but those involved in the Howard University human genome research. All uses of "African American" and "Caucasian" should be understood as categories uses by those proposing the research.

207. It is important to note that the ability of Dunston or other African American scientists and health-care professionals to speak for "African Americans" raises many complex questions. These questions will be addressed later in this chapter. For now, I seek to only explain their criticisms of the Diversity Project, and the notion of participation it incorporated.

208. Organizers of the Project invited a statistical geneticist Dunston worked with to the first planning workshop at Stanford in July of 1992. However, this investigator was out of the country for several months and so Dunston (who was handling the mail for the project) wrote back to ask if she could substitute. She received a positive response. Yet although in attendance, she did not play an active role in the workshop. She had no training in modeling genetic diversity (the subject of the workshop), and so mainly observed. Project organizers would later invite her to subsequent planning meetings, some of which she attended, some of which she could not attend. Dunston would become a supporter of the Diversity Project, whose goals she understood to be in part similar to those of the G-RAP project:

> I had a very similar argument to the scientific community—to the world if you will—that they were making for . . . the world population. And my argument was that the sampling that has been used for the Genome Project may be satisfactory to establish the map . . . but it will not be satisfactory to understand variation in the map. You must have the inclusion of the populations that reflect the variation. (Interview 1999)

As noted, however, one critical difference between G-RAP and the Diversity Project was that G-RAP proposed to characterize patterns of variation in African American genomes; to begin with, the Diversity Project did not.

209. This letter was added to the record of the Senate hearing on the Diversity Project held in April 1993.

210. Note that formulations like "these people in *our* population" and "*our* citizens" mark this formulation as part of a particular multicultural discourse [italics added]. The question would soon become who does this "our" refer to, and what did it reveal about the frame of reference employed by Project organizers?

211. Weiss's statement also represents a change in how Project organizers framed studying disease. From the beginning, they recognized that disease research would attract funders. Here it also acts as a potential answer to accusations that the Project would promote racism.

212. Weiss also argued that this sampling initiative (which in his mind was the Diversity Project) would, with "its inherent, social, historical and ethnic interest . . . help preserve for American science a semblance of scientific inquiry that is rapidly being lost amid the ever more blatant scramble for funding to do immediately 'relevant' work" (Weiss 1993, 44).

213. An exploration of how the ABGP came to use the term "descendant or culturally affiliated community," and what they took it to mean, would help shed further light on the process by which "people" and "communities" come to be defined as organizers of projects like the ABGP create processes for participation in genetic research.

214. In his account of the project, Blakey uses the term "descendant or culturally affiliated community" interchangeably with "African American public."

215. See chapter 4 for a description and exploration of the Wenner-Gren meeting.

216. Another Diversity Project organizer confirmed that before Mt. Kisco he and his colleagues were not working with a notion of partnership. He explains: "At that stage, even in Alghero [the site of the Sardinia meeting that took place in September 1993], we weren't that focused on treating the subjects of the research as active participants and partners, and I think that was a big mistake in the beginning and we should have seen that" (Interview A4).

217. For a description of the Model Ethical Protocol, see chapter 5.

218. This model of enrolling members of identified populations to sample themselves has become a popular strategy for dealing with the ethical problems raised by population-based research. This approach raises many problems that are beyond the scope of this chapter. To begin with, it shifts the ethical burden from the main researchers to the ethnic researchers they have enrolled to do the sampling. The Harvard/Millennium research collection effort in China is one such example (Pomfret and Nelson 2000). A detailed analysis of the problems of such an approach is needed.

219. This is not to say that Blakey's efforts did not raise their own problems. For example, as we saw in chapter 4, Blakey holds African Americans to be a coherent already constituted group, a group that could just be included in research. He does not explicitly address questions about the identity of African Americans, and complex questions about who (if anyone) can speak for a group called African American.

220. They do state that the "very point of the HGDP . . . is to collect information about communities that can be useful in improving understanding of the history and health of that community" (1472). However, the question of whether this understanding is of interest to the community is never raised.

221. Dukepoo stopped conducting his own genetic research shortly after learning about the Diversity Project. He believed that before conducting further research in Native communities a thorough discussion and debate of the issues needed to occur. As he explained in an article in the *San Francisco Chronicle*: "Who I am, and [who] my people were [were] too important to go on with that research [on the 'albino gene'] . . . I have a more important role to play" (Petit 1998). Dukepoo died an untimely death in 1999. Many Native as well as non-Native scientists considered him a credible spokesperson for tribes in the United States (Interview G, Interview J). He had been involved in the debates about the Project since 1993.

222. It is important to note here that for many members of tribes in the United States, colonization is not just a part of their history, it is a continuing process.

223. In addition to differences Dukepoo here notes between American Indians and African Americans, differences existed within each of these groups. I address these differences later in this section.

224. Others might challenge this notion that formal exploitation ended.

225. For an incisive and often-cited critique of this scientific research, see Deloria 1969.

226. See chapter 5 for a full explanation of this MacArthur funding. Most of the participants at this meeting were MacArthur Foundation grantees.

227. For scientists, on the other hand, the Out-of-Africa theory was just the beginning of what genetics could explore. One might even say that the theory that humans originated in Africa was the biologists' Big Bang; a whole universe of migrating and mixing genes was waiting to be explored. Note: In physics, the Big Bang theory is a theory of the origin of the universe.

228. The Indigenous Peoples Council on Biocolonialism formed in the mid-1990s in response to concerns created by the Human Genome Diversity Project (at that time, the group was known as the Indigenous Peoples Coalition against Biopiracy). It now has extended its concerns to include "the protection of [indigenous peoples'] genetic resources, indigenous knowledge, cultural and human rights from the negative effects of biotechnology," *www.ipcb.org* (accessed 12 June 2002). For concerns about the use of a genetic confirmation of the Bering Strait theory, also see Schmidt 2001, A218.

229. Many misinterpreted the legislation as saying that the state would *require* DNA testing to determine tribal identity. In fact, the legislator who proposed the bill intended that DNA tests be voluntarily requested by individuals as a way to build their case that they were Native American. As Kimberly Tallbear astutely explains, despite the "relatively benign intent" of the legislator, the bill does accept the notion that "biology can determine who rightly claims political and cultural authority," and "may be a forewarning of future laws and policies based on assumptions that a person's or a people's political rights and cultural identity are biologically determined" (Tallbear 2003, 85–86).

230. The extensive problems created by Amer-European attempts to determine the identity and origins of Native peoples cannot be adequately addressed in this book. For a powerful and incisive critique, see Deloria 1969.

231. And again, the prior question of who could speak for the people on these issues had not been resolved, or even discussed in any substantive way. Mooney had raised the issue in his letters to Greely in the spring of 1993, but it never became a focal concern of Diversity Project organizers—even among those involved in writing the Model Ethical Protocol.

232. Health had first been added to the Diversity Project's agenda in the context of trying to attract funding agencies (see chapter 4). Organizers recognized that health would be a bigger sell than origin stories. Health is now widely used to justify population genetics projects. See, for example, Pomfret and Nelson 2000.

233. Some also worried that pharmaceutical companies were the primary actors interested in disease research. Indeed, one participant in the MacArthur meeting reportedly argued that the research would be about wealth, not health.

234. Indeed, one organizer recognized: "To do the full medical things would cost like the Human Genome Project in each case" (Interview B). See also Marks 1998.

235. This language, "the facts," is that of one organizer of the meeting. As she explained: "You've got to get the facts out there first. You can't come in and say at a meeting, the first international meeting on sampling, the first meeting on sampling period, and say sampling is wrong when you don't even know what sampling is" (Interview H).

236. The meeting had originally caught the attention of those at the CBD meeting when it became known that the Diversity Project organizer, Luca Cavalli-Sforza, would speak. As one conference organizer recalled, these members of indigenous and public interest groups argued that the "Diversity Project coming to Canada" would promote biocolonialism and exploit Native peoples (Interview H).

237. This was the rate I paid. I arrived on the second day and entered the conference with ease. No trace of protest remained.

238. Many faulted Diversity Project organizers for not making more efforts to meet with Native peoples. After all, they argued, this is what the Project had received funding from the MacArthur Foundation to do. For their part, Diversity Project organizers argued that they had not met more often because there was no project and organizers were in a holding pattern waiting for the findings of the NRC committee. The meeting did follow directly in the wake of the release of the NRC report.

239. Greely later expressed frustration that Native peoples who critiqued the Project never really engaged with him in his effort to address their concerns through the Model Ethical Protocol's group consent provision (Interview 1999).

240. There were notable exceptions. Some invited to the meeting were willing to engage, but the legitimacy of their voices had been compromised by their other commitments: membership on NAmC committee; research position at the NIH.

241. For a discussion of the factors shaping answers to this question, see Benjamin 2000.

242. Which is not to say that they believed that genetic research could under no circumstances help Native peoples. Other human genetic variation projects have also raised grave concerns, but even the harshest critics hold out the possibility of designing genetic research that might help their communities. See, for example, Benjamin 2000.

243. Later organizers recognized that some populations might not want to participate (Interview A4, Interview R).

244. See chapter 5 for details.

245. Indeed, Diversity Project organizers had contacted this lawyer very early on. He did not take up the offer to get involved then because he did not have the time and the Project was not offering the kind of support he thought it would take to really consult with tribes. He only got involved once he became convinced the research posed a real "threat" to tribes.

246. For a description of this meeting, see chapter 5.

247. For an example of the way in which voices for groups are constructed and rendered a technical resource for political action, see Barrig 1999.

248. These controversies broke out at the same time that the Sawyer hearings were under way in the House Subcommittee on Census and Statistics and Postal Personnel. These hearings opened what would become the most intensive scrutiny of the definition and status of racial categories since the 1960s (Wright 1994, 46; Williams 2001). These interrogations linked to broader debates across the country about how people in different communities should identify themselves: As a member of one officially defined race? As a member of multiple races and ethnicities? When and where was it desirable to claim a racial or ethnic identity? When do we want to claim we are the same? When do we want to say we are different? To be successful, Diversity Project organizers also would have to successfully negotiate answers to these questions.

249. See chapter 3 for an analysis of Cavalli-Sforza's different arguments about race.

Chapter 7 Conclusion

250. CEPH is located in Paris, France, and was the first repository of human cell lines. For an account of CEPH see Rabinow 1999. In collaboration with CEPH, the Diversity Project has made available 1,064 cell lines made from the human tissues of individuals from populations spread across the globe (Cann et al. 2002, Cavalli-Sforza 2004). Although most of the cell lines are made from samples collected by others in the past and "donated" to the collaboration, some of the samples come from "HGDP collections in China and Southwest Asia." At least one proponent of the Diversity Project believes that this collection "could be the kernel for the international repository and database that make up so much of the HGDP's plans" (Greely 2001). In addition to the CEPH collaboration, today there are a variety of initiatives that are attempting to revive parts of the Diversity Project's original goals: collect samples from indigenous populations for the purposes of studying human origins and the history of human migrations. See, for example, the Africa Genome Initiative (http://www.africagenome.co.za). How-

ever, at the moment it seems unlikely that any of these initiatives will become as significant as proponents of the Diversity Project first hoped.

251. There are by now many other such projects, including in the United States the National Human Genome Research Institutes' Single Nucleotide Polymorphism Map Project, and the National Institutes of Environmental Health Sciences' Environmental Genome Project. Most recently, leaders at NHGRI, the Whitehead Institute in Cambridge, Massachusetts, and the Sanger Institute in the U.K. are leading an effort to create a haplotype map, a new kind of genome map designed to locate human genetic differences. For articles documenting some of these new initiatives see Marshall 1997, Nickerson 1998, Collins, et al. 1998, Sharp and Foster 2002, Couzin 2002.

252. In the last few years, those attempting to accommodate these new projects (such as the Environmental Genome Project) have increasingly begun to recognize that to move forward, their projects must address problems similar to those raised by the Diversity Project (Sharp 2002, Interview S).

253. Very much like the Diversity Project, the PDR seeks to represent human genetic diversity (Collins et al., 1998, http://www.nhgri.nih.gov/Grant_info/Funding/RFA/discover_polymorphisms.html). Unlike the Diversity Project, the PDR sought to represent just the human genetic diversity of the United States, and not the whole world.

254. NHGRI equated stripping these samples of their racial and ethnic markers with anonymizing them. Members of the "identified populations" in question quickly pointed out that researchers could easily re-identify the "racial and ethnic origins" of a sample by testing it for markers held to be unique to different racial and ethnic groups (Interview S, Interview G). This episode illustrates one of the many dilemmas that remain unresolved.

255. See chapter 3 for examples.

256. As a researcher, I did not escape this particular sorting process. It was important to many I interviewed to establish that I was a friend of the Project.

257. I explore this case in detail in chapter 5.

258. For a description of the Howard Center, see also *http://www.genomecenter.howard.edu.*

259. For meditations on these questions in the realm of state-centered approaches to inclusion, see Epstein 2003 and Brown 1995.

260. See chapter 1 for examples of these kinds of reductionist theories, as well as a more detailed discussion of the problems they create.

261. See chapter 5 for examples.

262. Among the many other efforts, of note for this book is what one organizer describes as a "morphed" Human Genome Diversity Project. Although this proponent of the initiative was willing to tell me of its existence, he was unwilling to provide any details. He justified his silence by stating the project had very little money, and it was still to early in the planning phase to say anything about it (Personal correspondence with author, November 2003).

263. Observations in this section are based on fieldwork and interviews with program directors and bioethicists involved in organizing human genetic variation research at the NIH (Interview S; Interview X; field notes from First Community

Consultation on the Responsible Collection and Use of Samples for Genetic Research, 25–26 September 2000, Bethesda, Maryland).

264. For a report of the "First Community Consultation on the Responsible Collection and Use of Samples for Genetic Research," see www.nigms.nih.gov/news/reports/community_consultation.html (accessed 13 June 2002).

265. This community consultation policy has itself generated many questions. To understand how issues raised by the Diversity Project are playing themselves out today, a critical analysis of this new institutional mechanism is needed.

Appendix A Methodological Appendix

266. Travel for this research was supported by grants from the Science and Technology Studies Department at Cornell University; the Cornell Ethical, Legal, and Ethical Issues Program; as well as a grant from the National Science Foundation, SBR-9818409. This support was absolutely crucial to the feasibility of this study.

267. Only one such interview was not recorded.

268. Often the consent process offered an opportunity to discuss the ethical and practical dilemmas created by the Diversity Project, as many of my subjects were themselves involved in debates about the meaning and practice of informed consent.

Bibliography

Akwasasne Notes. 1978. *Basic Call to Consciousness* (Summertown, TN: Book Publishing Company).

Akwasasne Task Force on the Environment, Research Advisory Committee. 1996. *Protocol for Review of Environmental and Scientific Research Proposals* (available from Akwesasne Task Force on the Environment, P.O. Box 992, Hogansburg, NY 13655).

Alvarado, Donna. 1995. " 'Race' A Useless Definition, Geneticists Says in Study." *Times-Picuyane* (21 February): A3.

Alvarez, Sonia E. 1997. "Latin American Feminisms 'Go Global': Trends of the 1990s and Challenges for the New Millenium." In *Culture of Politics/Politics of Cultures: Re-Visioning Latin American Social Movements*. Ed. by Sonia E. Alvarez, Evelina Dagnino, and Arturo Escobar (Boulder, CO: Westview Press): 293–324.

Amit-Talai, Vered. 1996. "The Minority Circuit: Identity Politics and the Professionalization of Ethnic Activism." In *Re-Situating Identities: The Politics of Race, Ethnicity and Culture*. Ed. by Vered Amit-Talai and Caroline Knowles (Ontario, Canada: Broadview Press): 89–114.

Amit-Talai, Vered, and Caroline Knowles, eds. 1996. *Re-Situating Identities: The Politics of Race, Ethnicity and Culture* (Ontario, Canada: Broadview Press).

Appiah, Anthony. 1990. "Racism." In *Anatomy of Racism*. Ed. by David Theo Goldberg (Minneapolis: University of Minnesota Press): 3–16.

Arditti, Rita. 1999. *Searching for Life: The Grandmothers of the the Plaza de Mayo and the Disappeared Children of Argentina* (Berkeley: University of California Press).

Asad, Talal, ed. 1973. *Anthropology and the Colonial Encounter* (New York: Humanities Press).

Baldwin, Kate. 1998. "Black Like Who? Cross-Testing the 'Real' Lines of John Howard Griffin's *Black Like Me*." *Cultural Critique*: 103–43.

Barkan, Elazar. 1992. *The Retreat of Scientific Racism* (Cambridge: Cambridge University Press).

Barnes, Barry, and David Bloor. 1982. "Relativism, Rationalism and the Sociology of Knowledge." In *Rationality and Relativism*. Ed. by Martin Hollis and Steven Lukes (Cambridge, MA: MIT Press): 21–47.

Barnes, Barry, and David Edge, eds. 1982. *Science in Context* (Cambridge, MA: MIT Press).

Barrig, Maruja. 1999. "The Persistence of Memory: Feminism and State in Peru in the 1990s." Paper presented to the Ford Foundation Comparative Regional Project on Civil Society and Democratic Governance in the Andes and Southern Core (November) http://www.ids.ac.uk/ids/civsoc/final/peru/per1.html (accessed 10 March 2002).

Bass, Thomas A. 1994. *Reinventing the Future: Conversations with the World's Leading Scientists* (New York: Addison-Wesley Publishing Company).

Beecher, Henry E. 1987. "Ethics and Clinical Research." *New England Journal of Medicine* 274: 1354–60.

Benedict, Ruth. 1940. *Race, Science and Politics* (New York: Modern Age Books).

Benjamin, Craig. 1996. "Check Your Ethics at the Door: Exclusion of Indigenous Peoples' *Representatives from DNA Sampling* Conference Casts Doubts on Gene Hunters' Sincerity." http://nativenet.uthscsa.edu/archive/nl/9609/0060.html (accessed 15 February 2001).

———. 2000. "Sampling Indigenous Blood for Whose Benefit?" *Native Americas* XVII (1): 38–45.

Billings, Paul R. 1992. "Discrimination as a Consequence of Genetic Testing." *American Journal of Human Genetics* 50: 476–82

Blakey, Michael. 1998. "Beyond European Enlightenment: Toward a Critical and Humanistic Biology." In *Building a New Biocultural Synthesis: Political-Economic Perspectives on Human Biology*. Ed. by Alan Goodman and Thomas Leatherman (Ann Arbor: University of Michigan Press): 379–406.

Boas, Franz. 1910. *Changes in Bodily Form of Descendants of Immigrants* (Washington DC: U.S. Immigration Commission).

Bowcock, Anne, and Luca Cavalli-Sforza. 1991. "The Study of Variation in the Human Genome." *Genomics* 11 (Summer): 491–98.

Bowker, Geoff, and Susan Leigh Star. 1999. *Sorting Things Out* (Cambridge, MA: MIT Press).

Boyd, William. 1950. *Genetics and the Races of Man* (Boston: Little, Brown and Company).

Brace, C. L. 1964. "On the Race Concept." *Current Anthropology* 5(4): 313–14.

Branch, Taylor. 1988. *Parting the Waters* (New York: Simon and Schuster).

Brown, Wendy. 1995. *States of Injury: Power and Freedom in Late Modernity* (Princeton, NJ: Princeton University Press).

Brown v. Board of Education. 1954. 347 U.S. 483.

Buffon, Georges Lois Leclerc. 1817. *A Natural History, General and Particular* (London: Richard Evans).

Burchard E., et al. 2003. "The Importance of Race and Ethnic Background in Biomedical Research." *NEJM* 348(12): 1170–75.

Butler, Declan. 1995. "Genetic Diversity Panel Fails to Impress International Ethics Panel." *Nature* 377 (5 October): 373.

Callon, Michel. 1999. "Some Elements of a Sociololgy of Translation: Domestication of the Scallops and the Fishermen of St. Brieuc Bay." In *The Science Studies Reader.* Ed. by Mario Biagioli (Routledge: New York, 1999): 67–83.

Cann, Howard. 1998. "Human Genome Diversity." *C. R. Acad. Sci. Paris, Sciences de la vie* 321: 443–46.

Cann, Howard, et al. 2002. "A Human Genome Diversity Cell Line Panel." *Science* 296: 261–62.

Cann, Rebecca, Mark Stoneking, and Allan C. Wilson. 1987. "Mitochondiral DNA and Human Evolution." *Nature* 325 (1 January): 31–36.

Cantor, Charlie, and Cassandra L. Smith. 1999. *Genomics: The Science and Technology behind the Human Genome Project* (New York: John Wiley and Sons).

Cavalli-Sforza, L. Luca. 1986. "Introduction." In *African Pygmies*. Ed. by Luca Cavalli-Sforza (London: Academic Press).

———. 1989. Letter to Robert Cook-Deegan (September 22).

———. 1990. "Opinion: How Can One Study Individual Variation for 3 Billion Nucleotides of the Human Genome?" *Am. J. Hum. Genet.* 46: 649–51.

———. 1991. "Genes, Peoples and Languages." *Scientific American* 265 (November): 104.

———. 1994. "The Human Genome Diversity Project." An Address Delivered to a Special Meeting of UNESCO. Paris, France (12 September).

———. 1998. "The Chinese Human Genome Diversity Project." *Proc. Natl. Acad. Sci. USA* 95: 11501–03.

———. 2000. *Genes, Peoples, and Languages*. Translation by Mark Seielstad (New York: North Point Press).

———. 2004. "Diversity Project Takes Time but Reaps Rewards." *Nature* 428: 467.

Cavalli-Sforza, L. Luca, and Walter Bodmer. 1999 (1971). *The Genetics of Human Populations* (New York: Dober Publications).

Cavalli-Sforza, L. Luca, and Francesco Cavalli-Sforza. 1995. *The Great Human Diasporas* (Reading, MA: Addison-Wesley Publishing Company).

Cavalli-Sforza, L. L., and Marcus, Feldman. 1981. *Cultural Transmission and Evolution, A Quantitative Approach* (Princeton, NJ: Princeton University Press).

Cavalli-Sforza, L. Luca, Paolo Menozzi, and Alberto Piazzi. 1994. *The History and Geography of Human Genes* (Princeton, NJ: Princeton University Press).

Cavalli-Sforza, L. Luca, Allan C. Wilson, Charles R. Cantor, Robert M. Cook-Deegan and Mary-Claire King. 1991. "Call for a Worldwide Survey of Human Genetic Diversity: A Vanishing Opportunity for the Human Genome Project." *Genomics* 11 (Summer): 490–91.

Cayuqueo, Nilo. 1993. "HGD Project, SAIIC Response." http://nativenet.uthscsa.edu/archive/nl/9311/0069.html (accessed 6 October 1999).

Churchill, Ward. 1994. *Indians Are Us? Culture and Genocide in Native North America* (Toronto: Between the Lines).

Clark, W. E. LeGros. 1964. *The Fossil Evidence for Human Evolution* (Chicago: Chicago University Press).

Clifford, James, and George E. Marcus, eds. 1986. *Writing Culture: The Poetics and the Politics of Ethnography* (Berkeley: University of California Press).

Clinton, William J. 2000. "Statement Transcribed on Decoding of Genome." *New York Times* (27 June): D8.

Coghian, Andy. 1998. "Selling the Family Secrets." *New Scientist* (5 December): 20–21.

Cohen, Felix. 1982. *Handbook of Federal Indian Law* (Charlottesville, VA: Michie Bobbs-Merrill Law Publishers).

Cohen, Jean. 1985. "Strategy or Identity: New Theoretical Paradigms and Contemporary Social Movements." *Social Research* 52: 663–716.

Cold Spring Harbor Symposium on Quantitative Biology. 1950. *Origin and Evolution of Man* (Cold Spring Harbor, NY: Long Island Biological Association).

Coleman, William. 1964. *Georges Cuvier, Zoologist: A Study in the History of Evolution Theory* (Cambridge: Harvard University Press).

Collins, Francis S., Lisa D. Brooks, and Aravinda Chakravarti. 1998. "A DNA Polymorphism Discovery Resource on Human Genetic Variation." *Genome Research* 8: 1229–31.

Collins, H. M. 1985. *Changing Order: Replication and Induction in Scientific Practice* (London: Sage).

"Commentary: The Urgent Need to Preserve DNA Resources from Isolated Human Populations." 1991. Draft of Commentary to be published in *Genomics*.

Committee on Governmental Affairs. 1993. *Human Genome Diversity Project: Hearing before the Committee on Governmental Affairs*, U.S. Senate, 103rd Congress 1st Session (26 April).

Cook-Deegan, Robert. 1991. Letter to Luca Cavalli-Sforza (12 July).

———. 1994. *The Gene Wars* (New York: W. W. Norton & Company).

Cook-Deegan, Robert, and James Watson. 1990. "The Human Genome Project and International Health." *JAMA* 263(24): 3322–24.

Coon, Carleton. 1964. "Comments." *Current Anthropology* 5(4): 314.

———. 1966. "The Taxonomy of Human Variation." *Annals of the New York Academy of Sciences* 134(2): 516–23.

Cooper, Richard S., Jay S. Kaufman, and Ryk Ward. 2003. "Race and Genomics." *NEJM* 348(12): 1166–70.

Count, Earl W. 1964. "Comments." *Current Anthropology* 5(4): 314–16.

Couzin, Jennifer. 2002. "New Mapping Project Splits the Community." *Science* 296 (24 May): 1391–93.

Cruse, Harold. 1968. *The Crisis of the Negro Intellectual* (New York: William Morrow).

Cultural Survival et al. 1996. *Open Letter to First International Conference on DNA Sampling* (Montreal, Canada).

Darwin, Charles. 1859. *On the Origin of Species by Means of Natural Selection, Or the Preservation of Favoured Races in the Struggle for Life* (London: Murray).

Daston, Lorraine, ed. 2000. *Biographies of Scientific Objects* (Chicago: University of Chicago Press).

Deloria, Vine Jr. 1969. "Anthropologists and Other Friends" in *Custer Died for Your Sins* (New York: Avon).

———. 1985. *American Indian Policy in the Twentieth Century* (Norman: University of Oklahoma Press).

Dezalay, Yvez, and Bryant G. Garth. 1996. *Dealing in Virtue: International Commercial Arbitration and the Construction of a Transnational Legal Order* (Chicago: University of Chicago Press).

Diamond, Jared M. 1993. *The Third Chimpanzee: The Evolution and Future of the Human Animal* (New York, NY: Harper Perennial).

"Diversity Project 'Does Not Merit Federal Funding.' " *Nature* 389 (23 October): 774.

DNA Learning Center. 1992. "Extending Our Expertise to Minority and Disadvantaged Settings." *DNA Learning Center Annual Report* (Cold Spring Harbor): 376–78.

Dobzhansky, Theodosius. 1937 (1951). *Genetics and the Origin of Species*, 3d ed. (New York: Columbia University Press).

———. 1950. "The Genetic Nature of Differences among Men." In *Evolutionary Thought in America*, Ed. by Stow Peters (New Haven: Yale University Press).

———. 1962. "Comment." *Current Anthropology* 3(3): 279–80.

———. 1963. "A Debatable Account of the Origin of Races." *Scientific American* 208: 169–72.

DuBois, W.E.B. 1961 (1903). *The Souls of Black Folk* (Greenwich, CT: Fawcett).

Dukepoo, Frank. 1997. "Native American Perspectives on Genetic Patenting." Presented to the American Association for the Advancement of Science, Washington D.C. (1 October).

———. 1998. "Commentary on "Scientific Limitations and Ethical Ramifications of a Non-representative Human Genome Project." *Science and Engineering Ethics* 4: 171–80.

———. 1998b. "Sensitivities and Concerns in Native American Communities." Hearing of the National Bioethics Advisory Commision (December 2). http://bioethics.gov/transcripts/jul98/native.html (accessed 12 June 2000).

Dulbecco, Renato. 1986. "A Turning Point in Cancer Research: Sequencing the Human Genome." *Science* 231: 1055–56.

Dunn, Leslie Clarence. 1959. *Heredity and Evolution in Human Populations* (Cambridge, MA: Harvard University Press).

Dunn, Leslie Clarence, and Theodosius Dobzhansky. 1946. *Heredity, Race and Society* (New York: Mentor Books).

Dunston, Georgia. 1998. "G-RAP: A Model HBCU Genomic Research and Training Program." In *Plain Talk about the Human Genome Project*. Ed. by Edward Smith and Walter Sapp (Tuskegee, AL: Tuskegee University Press).

Duster, Troy. 1990. *Backdoor to Eugenics* (London: Routledge).

Dyer, Allen R. 1982. "The Dynamics of Dependency Relationships." *IRB: A Review of Human Subjects Research* 4(7).

Eagleton, Terry. 1991. *Ideology: An Introduction* (New York: Verso).

Engels, Frederick. 1937 [1893]. "Letter to Mehring." In *Karl Marx and Friedrich Engels, Selected Correspondence, 1846–95*, ed. Dona Torr (New York: International Publishers): 510–12.

Epstein, Steven. 1991. "Democratic Science? AIDS Activism and the Contested Construction of Knowledge." *Socialist Review* (April–June): 35–64.

———. 1996. *Impure Science* (Berkeley: University of California Press).

———. 2000. "Medical Subjects and Objects: Emergent Lesbian and Gay Male Health Agendas on the National Stage." Paper presented to the American Sociological Association, Anaheim, California (14 August).

———. 2003. "Sexualizing Governance and Medicalizing Identities: The Emergence of 'State-Centered' LGBT Health Policies in the United States." *Sexualities* 6(2): 131–71.

———. Forthcoming. "Bodily Differences and Collective Identities: The Politics of Gender and Race in Biomedical Research in the United States." *Body and Society.*

Ezrahi, Yaron. 1990. *The Descent of Icarus* (Cambridge: Harvard University Press).

Feldman, Marcus, and Luca Cavalli-Sforza. 1981. *Cultural Transmission and Evolution: A Quantitative Approach* (Princeton, NJ: Princeton University Press).

Fields, Barbara. 1990. "Slavery, Race and Ideology in the United States of America." *New Left Review* 181 (May/June): 95–188.

First International Conference on DNA Sampling. 1996. *Conference Program.* Organized by Centre de recherché en droit public, Universite de Montreal, and the Health Law Institute, University of Alberta.

Flint, Anthony. 1995. "Don't Classify by Race, Urge Scientists." *Boston Globe* (5 March): B1.

Foucault, Michel. 1966. *The Order of Things: An Archaeology of the Human Sciences* (New York: Vintage).

———. 1972. *The Archaeology of Knowledge* (New York: Pantheon Books).

———. 1973. *The Birth of the Clinic: An Archaeology of Medical Perception* (New York: Vintage Books).

———. 1975. *Discipline and Punish: The Birth of the Prison* (London: Allen Lane).

———. 1976. *The History of Sexuality, Volume I: An Introduction* (London: Allen Lane).

———. 1977. "Nietzsche, Genealogy, History." In *Language, Counter-Memory, Practice.* Ed. by Donald F. Bouchard (Ithaca, NY: Cornell University Press).

———. 1980. "Truth and Power." In *Power/Knowledge: Selected Readings and Other Writings 1972–1977.* Ed. by Colin Gordon (New York: Pantheon Books): 109–33.

Freeman, Harold P. 1998. "The Meaning of Race in Science—Considerations for Cancer Research." *Cancer* 82(1): 219–25.

Fried, Morton. 1975. *The Notion of Tribe* (Menlo Park, CA: Cummings Publication Co.).

Fullilove, Mindy Thompson. 1998. "Comment: Abandoning 'Race' as a Variable in Public Health Research—An Idea Whose Time Has Come." *American Journal of Public Health* 88(9): 1297–98.

Gannett, Lisa. 1999. "Theodosius Dobzhansky and the Typological-Population Distinction." A paper presented to the International Society for the History, Philosophy and Social Studies of Biology, Oaxaca, Mexico (8 July).

———. 2001. "Scientific Investigations of Human Genome Diversity: Methodological, Theoretical and Ethical Limits of 'Population Thinking.' " *Philosophy of Science* 68(3): 479–92.

Garn, Stanley M. 1964. "Comments." *Current Anthropology* 5(4): 316.
Gates, Henry Louis Jr., ed. 1986. *"Race," Writing, and Difference* (Chicago: University of Chicago Press).
Gibbons, Michael. 1994. *The New Production of Knowledge: The Dynamics of Science and Research in Contemporary Societies* (London; Thousand Oaks, CA: Sage Publications).
Gieryn, Thomas F. 1983. "Boundary-Work and the Demarcation of Science from Non-Science: Strains and Interests in Professional Ideologies of Scientists." *American Sociological Review* 48: 781–95.
Gilroy, Paul. 2000. *Beyond Camps: Race, Identity and Nationalism at the End of the Colour Line* (London, England: The Penguin Press).
Glass, Bentley. 1986. "Genetics Embattled: Their Struggle against Rampant Eugenics and Racism in America during the 1920s and 1930s." In *Proceedings of the American Philosophical Society* 130: 130–54.
Gobert, Judy. 1998. Letter to North American Committee of the Human Genome Diversity Project. (9 October).
———. 1998 "Let's Walk on the Wild Side!" A paper presented to the The First North American Conference on Genetics and Native Peoples, Polson, MT (11–12 October).
Goldberg, David Theo. 1993. *Racist Culture: Philosophy and the Politics of Meaning* (Cambridge, MA: Blackwell).
Goodman, Alan. 1992. Letter to Leslie Roberts (18 December)
———. 1997. "Bred in the Bones?" *The Sciences* (March/April): 20–25.
Goodman, Alan, and Evelynn Hammonds. 2000. "Reconciling Race and Human Adaptability: Carleton Coon and the Persistence of Race in Scientific Discourse." In *Kroeber Anthropological Society Papers*. Ed. by Jonathan Marks (Berkeley: University of California Press): 28–43.
Goodman, Nelson. 1978. *Ways of Worldmaking* (Indianapolis: Hackett).
Goodman, Alan, and Thomas Leatherman, eds. 1998. *Building a New Biocultural Synthesis* (Ann Arbor: University of Michigan).
Gootman, Elissa. 2003. "DNA Evidence Frees 3 Men in 1984 Murder of L.I. Girl." *New York Times* (12 June): B1.
Graves, Joseph. 2001. *The Emperor's New Clothes: Biological Theories of Race at the Millennium* (New Brunswick, NJ: Rutgers University Press).
Greely, Henry T. 1993b. Letter to Pat Mooney (8 June).
———. 1993c. Letter to Pat Mooney (8 July).
———. 1993d. Letter to Pat Mooney (1 November).
———. 1993e. "The HGD Project—a response to SAIIC" http://nativenet.uthscsa.edu/archive/nl/9307/0046.html (accessed 6 October 1999).
———. 1993f. "Re: DNA Probe Angers the 'Endangered.' " http://nativenet.uthscsa.edu/archive/nl/9307/0046.html (accessed 6 October 1999).
———. 2001. "Human Genome Diversity: What about the Other Human Genome Project?" *Genetics* (March): 222–27.
Greely, Henry T., and Luca Cavalli-Sforza. 1993. "Human Genome Diversity Project—Organizers' Response." http://nativenet.uthscsa.edu/archive/nl/9307/0046.html (accessed 6 October 1999).

Greenberg, Judith. 2000. "Special Oversight Groups to Add Protections for Population-Based Repository Samples." *Am. J. Hum. Genet.* 66: 745–47.

Guerrero, M.A.J. 1998. "Eugenics Coding and American Racisms: Focus on the Human Genome Diversity Project as 'The Great Spiritual Ripoff.' " (Paper presented to the First North American Conference on Genetics and Native Peoples, Polson, MT (11–12 October).

Guillaumin, Collette. 1988. "Race and Nature and the System of Marks: The Idea of a Natural Group and Social Relationships." *Feminist Issues* Fall: 25–42.

Gusfield, Joseph. 1981. *The Culture of Public Problems: Drinking-Driving and the Symbolic Order* (Chicago: University of Chicago Press).

Habermas, J. 1975. *Legitimation Crisis* (Boston: Beacon Press).

Hacking, Ian. 1999. *The Social Construction of What?* (Cambridge, MA: Harvard University Press).

Hammonds, Evelynn. 1997. "New Technologies of Race." In *Processed Lives: Gender and Technology in Everyday Life*. Ed. by J. Terry and M. Calvert (New York: Routledge): 107–22.

Hansen, Roger D. 1979. *Beyond the North-South Stalemate* (New York: McGraw-Hill Book Company).

Haraway, Donna. 1989. *Primate Visions* (New York: Routledge).

———. 1991. *Simian, Cyborgs, and Women* (New York: Routledge).

———. 1996. "Modest Witness: Feminist Diffractions in Science Studies. In *The Disunity of Science: Boundaries, Contexts, and Power.* Ed. by Peter Galison and David J. Stump (Stanford, CA: Stanford University Press): 428–42.

———.1997. *Modest_Witness@Second_Millenium.FemaleMan©_Meets_Oncomouse™* (New York: Routledge).

Harrison, Ira, and Faye Harrison. 1991. *African American Pioneers in Anthropology* (Urbana: University of Illinois Press).

Harry, Debra. 1995. "Patenting of Life and Its Implications for Indigenous Peoples." *Information about Intellectual Property Rights* 7: 1–2.

Hayden, Corrine P. 1998. "A Biodiversity Sampler for the Millenium." In *Reproducing Reproduction: Kinship, Power, and Technological Innovation*. Ed. by Sarah Franklin and Helen Ragone (Philadelphia: University of Pennsylvania Press): 173–206.

Herrnstein, Richard, and Charles Murray. 1994. *The Bell Curve* (New York: The Free Press).

Herzig, Rebecca. 1999. "Removing Roots: 'North American Hiroshima Maidens' and the X Ray." *Technology and Culture* 40(4): 723–45.

Higginbotham, Evelyn Brooks. 1992. "African-American Women's History and the Metalanguage of Race." *Signs: A Journal of Women in Culture and Society* 17: 251–74.

Hilgartner, Stephen. 2000. *Science on Stage: Expert Advice as Public Drama* (Stanford: Stanford University Press).

———. 2002. "Biotechnology." *International Encyclopedia of the Social & Behavioral* Sciences, 2nd ed. (New York: Elsevier): 1235–40.

Horowitz, Irving L., ed. 1967. *The Rise and Fall of Project Camelot: Studies in the Relationship between Social Science and Practical Politics* (Cambridge: MIT Press).

Hotz, Robert Lee. "Scientists Say Race Has No Biological Basis." *Los Angeles Times* (February 20): A1.

Hull, David L. 1965. "The Effect of Essentialism on Taxonomy: Two Thousand Years of Stasis." *British Journal for the Philosophy of Science* 15: 314–26.

Human Genome Diversity Project. 1992a. *Human Genome Diversity Workshop 1* (Stanford, California: Stanford).

———. 1992b. *Human Genome Diversity Workshop* 2 (State College: Penn State University).

———. 1993. *Human Genome Diversity Project—Summary of Planning Workshop 3(B): Ethical and Human Rights Implications* (16–18 February).

Human Genome Organization. 1993. *The Human Genome Diversity (HGD) Project—Summary Document.* Report of the Intenational Planning Workshop held in Porte Conte, Sardinia, Italy (9–12 September).

Huxley, Julian. 1964. "Comments." *Current Anthropology* 5(4): 316–17.

Huxley, Julian, and Alfred Haddon. 1936. *We Europeans: A Survey of "Racial Problems"* (New York: Harper).

Indigenous Peoples Council on Biocolonialism. 1998. "Resolution by Indigenous Peoples." http://www.ipcb.org/resolutions/index.htm (accessed 5 March 2002).

———. 1999. *A Primer and Resource Guide.* www.ipcb.org (accessed 5 March 2002).

———. 2000. *Indigenous Research Protection Act.* http://www.ipcb.org/pub/irpaintro.html (accessed 4 March 4 2002).

———. 2001. "Genetic 'Markers'—Not a Valid Test of Native Identity." http://www.ipcb.org/pub/index.htm (accessed 10 March 2002).

Jackson, Fatimah. 1998. "Scientific Limitations and Ethical Ramifications of a Non-representative Human Genome Project: African American responses." *Science and Engineering Ethics* 4: 155–70.

Jackson, John P. Jr. 2001. " 'In Ways Unacademical': The Reception of Carleton Coon's *The Origin of Races." Journal for the History of Biology* 34:247–85.

Jasanoff, J., G. E. Markle, J. C. Peterson, T. Pinch, eds. 1995. *Handbook of Science and Technology Studies* (Thousand Oaks, CA: Sage).

Jasanoff, Sheila. 1990. *The Fifth Branch: Science Advisers as Policymakers* (Cambridge, MA: Harvard University Press).

———. 1992. "Science, Politics, and the Renegotiation of Expertise at the EPA." *Osiris* 7: 195–217.

———. 1995. *Science at the Bar: Law, Science and Technology in America* (Cambridge, MA: Harvard University Press).

———. 2002. "The Life Sciences and the Rule of Law." *Journal of Molecular Biology* 319(4): 891–99.

———. 2004. "Ordering Knowledge, Ordering Society." In *States of Knowledge: The Co-Production of Science and Social Order.* Ed. by Sheila Jasanoff (London: Routledge).

———. 2004b. "Afterword." In *States of Knowledge: The Co-Production of Science and Social Order.* Ed. by Sheila Jasanoff (London: Routledge).

Jensen, Arthur R. 1969. "How Much Can We Boost IQ and Scholastic Achievement?" *Harvard Educational Review,* 39(1): 1–123.

John Moore v. Regents of the University of California et al., 241 Cal. Rptr. 147 (1990).

Johnston, Francis E. 1966. "The Population Approach to the Human Variation." *Annals of the New York Academy of Sciences* 134(2): 507–15.

Jones, James H. 1981. *Bad Blood: The Tuskegee Syphilis Experiment* (New York: The Free Press).

Jones, Steve. 1993. *The Language of the Genes* (New York: Anchor Books).

Juengst, Eric. 1995. "Lineage, Land Tenure, and Demic Discrimination: The Social Risks of the Human Genome Diversity Project," paper presented at A Public Symposium on Genetics and the Human Genome Project: Where Scientific Cultures and Public Cultures Meet, Stanford University (3–4 November).

———. 1998. "Group Identity and Human Diversity: Keeping Biology Straight from Culture." *Am. J. Hum. Genet.* 63: 673–77.

Kaiser, Jocelyn. 2000. "Higher Profile for Minority Health." *Science* 290: 1667–68.

Kaufmann, Alain. 2001. "Mapping at the Genethon Laboratory: The French Muscular Distrophy Association and the Politics of the Gene." A paper presented to The Mapping Cultures of Twentieth-Century Genetics, Max-Plack Institute, Berlin, Germany (1–4 March).

Kay, Lily E. 1993. *The Molecularization of Life: Caltech, the Rockefeller Foundation, and the Rise of the New Biology* (New York: Oxford University Press).

Keller, Evelyn Fox. 1985. *Reflections on Gender and Science* (New Haven: Yale University Press).

———. 1992. "Fractured Images of Science, Language and Power." In *Essays on Language, Gender and Science* (New York: Routledge): 93–110.

———. 2000. *The Century of the Gene* (Cambridge, MA, and London: Harvard University Press).

Kevles, Daniel. 1985. *In the Name of Eugenics: Genetics and the Uses of Human Heredity* (Berkeley: University of California Press).

Kevles, Daniel, and Leroy Hood, eds. 1992. *The Code of Codes: Scientific and Social Issues in the Human Genome Project* (Cambridge, MA: Harvard University Press).

Kidd, Kenneth, et al. 1992. "Forum on Human Genetic Diversity." *Human Biology* 65:1: 7–9.

King, Mary-Claire. 1996. "Relevance of the Human Genome Diversity Project to Biomedical Research." Statement submitted to the National Academy of Sciences committee charged with evaluating the Human Genome Diversity Project. http://www.stanford.edu/group/morrinst/hgdp/MCK2NRC.html (accessed 11 June 2001).

King, Mary-Claire, and Allan Wilson. 1975. "Evolution at Two Levels in Humans and Chimpanzees." *Science* 188 (April 11): 107–16.

Kitcher, Phillip. 2001. *Science, Truth, and Democracy* (Oxford: Oxford University Press).

Kleinman, Arthur, Renee Fox, and Allan Brandt. 1999. "Introduction: Bioethics and Beyond." *Daedalus: Journal of the American Academy of Arts and Sciences* 128(4): vii–x.

Kleinman, Daniel Lee, ed. 2000. *Science, Technology, and Democracy* (Albany: State University of New York).

Kloppenburg, Jack. 1988. *First the Seed: The Political Economy of Plant Biotechnology* (Cambridge, UK: Cambridge University Press).

Kluger, Richard. 1975. *Simple Justice: The History of* Brown v. Board of Education *and Black America's Struggle for Equality* (New York: Vintage Books).

Kuklick, Henrika. 1991. *The Savage Within: The Social History of British Anthropology, 1885–1945* (Cambridge: Cambridge University Press).

Kull, Andrew. 1992. *The Color-Blind Constitution* (Cambridge, MA: Harvard University Press).

Kuznick, Peter. 1987. *Beyond the Laboratory: Scientists as Political Activists in 1930s America* (Chicago: Chicago University Press).

LaCapra, Dominick. 1994. *Representing the Holocaust: History, Theory, Trauma* (Cornell University Press: Ithaca).

Latour, Bruno. 1991. *We Have Never Been Modern* (Cambridge, MA: Harvard University Press).

———. 1992. "The Sociology of a Few Mundane Artifacts." In *Shaping Technology/Building Society.* Ed. by Wiebe Bijker and John Law (Cambridge, MA: MIT Press): 225–58.

Latour, Bruno, and Woolgar, Steve. 1986. *Laboratory Life* (Princeton, NJ: Princeton University Press).

Lewin, Roger. 1982. *Thread of Life* (New York: W. W. Norton and Company).

———. 1993. "Genes from a Disappearing World." *New Scientist* (May 29): 25–29.

Lewontin, Richard. 1972. "The Apportionment of Human Diversity." *Evol. Biol.* 6: 381–98.

Lippman, Abby. 1991. "Prenatal Genetic Testing and Screening: Constructing Needs and Reinforcing Inequalities." *American Journal of Law and Medicine* 17: 15–50.

Livingstone, Frank. 1962. "On the Non-Existence of Human Races." *Current Anthropology* 3(3): 279.

Lock, Margaret. 1994. "Interrogating the Human Diversity Project." *Soc. Sco. Med.* 39(5): 603–606.

Longino, Helen. 1990. *Science as Social Knowledge* (Princeton: Princeton University Press).

MacIlwain, Collin. 1997. "Diversity Project Does Not Merit Federal Funding." *Nature* 389 (23 October): 774.

Majumder, Partha. 2000. "Lessons Learned from the Indian Experience." Talk presented to the First Community Consultation on the Responsible Collection and Use of Samples for Genetic Research, Bethesda, Maryland (September 25–26).

Malcolm X and Alex Haley. 1964. *The Autobiography of Malcolm X* (New York: Ballantine Books).

Marcus, George. 1998. *Ethnography through Thick and Thin* (Princeton: Princeton University Press).

Marcus, George, ed. 1998. *Corporate Futures: The Diffusion of the Culturally Sensitive Corporate Form* (Chicago: The University of Chicago Press).

Marcus, George, and Michael Fischer. 1986. *Anthropology as Cultural Critique: An Experimental Moment in the Human Sciences* (Chicago: The University of Chicago Press).

Marks, Jonathan. 1995. "The Human Genome Diversity Project: Good for if Not Good as Anthropology?" *Anthropology Newsletter* (April): 72.

———. 1996. "The Legacy of Serological Studies in American Physical Anthropology." *Hist. Phil. Life. Sci.* 18: 345–62.

Marks, Jonathan. 1998. "Letter: The Trouble with the HGDP." *Molecular Medicine Today* (June): 243.

Marks, Jonathon, ed. 2000. *Kroeber Anthropological Society Papers* (Berkeley: University of California Press).

Marshall, Marshall. 1997. "Gene Prospecting in Remote Populations." *Science* 278 (24 October): 565.

Mayr, Ernst. 1980. "The Role of Systematics in the Evolutionary Synthesis." In *The Evolutionary Sythesis: Perspectives on the Unification of Biology.* Ed. by William Provine and Ernst Mayr (Cambridge, MA: Harvard University Press).

Mead, Aroha Te Pareake. 1996. "Genealogy, Sacredness, and the Commodities Market." *Cultural Survival Quarterly* (Summer): 46–50.

Melucci, Alberto. 1989. *Nomads of the Present: Social Movements and Individual Needs in Contemporary Society.* Ed. by John Keane and Paul Mier (Philadelphia: Temple University Press).

Merton, Robert. 1942. *The Sociology of Science* (Chicago: University of Chicago Press).

Mittman, Ilana Suez. 1997. "Commentary: We Must All Be Equal Partners in the New Age of Genetics." *The Scientist* 11(20): 8.

Montagu, Ashley. 1942. *Man's Most Dangerous Myth: The Fallacy of Race* (New York: Columbia University Press).

———. 1964. "Comments." *Current Anthropology* 5(4): 317.

———. 1972. *Statement on Race* (New York: Oxford University Press).

Montoya, Michael. 2001. "Bioethnic Conscriptions: Biological Capital and the Genetics of Type 2 Diabetes." A paper presented at the American Sociological Association Annual Meeting, Anaheim, California (14 August).

Mooney, Pat. 1993. Letter to Henry T. Greely (2 June).

———. 1993b. Letter to Henry T. Greely (30 June).

———. 1993c. Letter to Henry T. Greely (2 November).

———. 1996. "The Parts of Life: Agricultural Biodiversity, Indigenous Knowledge, and the Role of the Third System." *Development Dialogue* 1–2:1–84.

Moore, John H. 1993. "Organizer's Statement: Anthropological Perspectives on the Human Genome Diversity Project." A statement presented to the Wenner-Gren conference, Anthropological Perspectives on the Human Genome Diversity Project, Sever Springs Center, Mt. Kisko, New York (3–7 November).

———. 1994. "Ethnogenetic Theory." *National Geographic Research and Exploration* 10(1): 10–23.

———. 1995. "The End of a Paradigm." *Current Anthropology* 36(3): 530–31.

———. 1996. "Is the Human Genome Diversity Project a Racist Enterprise." In *Race and Other Misadventures: Essays in Honor of Ashley Montagu in His Ninetieth Year* (Dix Hills, New York: General Hall): 217–29.

Morgan, Thomas Hunt. 1932. *The Scientific Basis of Evolution* (New York: W. W. Norton).

Murphy, Michelle. 1998. *Sick Buildings and Sick Bodies: The Materialization of an Occupational Illness in Late Capitalism.* Thesis (Ph.D). Harvard University (Cambridge, MA).

Myrdal, Gunnar. 1962 (1944). *An American Dilemma: The Negro Problem and Modern Democracy* (New York: Harper & Row).

National Commission for the Protection of Human Subjects of Biomedical and Behavioral Research. 1979. *The Belmont Report: Ethical Principles and Guidelines for the Protection of Human Subjects of Research.* Washington, DC: Office for Protection from Research Risks, National Institutes of Health, Dept. of Health, Education, and Welfare (18 April).

National Institutes of Health. 1994. NIH Guidelines on the Inclusion of Women and Minorities as Subjects in Clinical Research. *Federal Register* 59 (Part VIII): 14508–13.

National Research Council. 1997. *Evaluating Human Genetic Diversity* (Washington DC: National Academy Press).

Native-L. 1993a. "Call to Stop Human Genome Project." http://nativenet.uthscsa.edu/archive/nl/9307/0046.html (accessed 6 October 1999).

———. 1993b. "Human Genome Diversity Project." http://nativenet.uthscsa.edu/archive/nl/9307/0046.html (accessed 6 October 1999).

Natowicz, M. R. 1992. "Genetic Discirimination and Law." *American Journal of Human Genetics* 50: 465–75.

Neel, James. 1994. *Physcian to the Gene Pool: Genetic Lessons and Other Stories* (New York: J. Wiley).

Nickens, H. 1993. "Minority Health Research Issues." *Science, Technology, & Human Values* 18: 506–10.

Nickerson, D. A., et al. 1998. "DNA Sequence Diversity in a 9.7-kb Region of the Human Lipoprotein Lipase Gene." *Nature Genetics* (19 July): 233–40.

Noble, David. 1984. *Forces of Production.* (New York : Knopf).

North American Regional Committee of the Human Genome Diversity Project. 1994c. "Answers to Frequently Asked Questions about the Human Genome Diversity Project." http://www.stanford.edu/group/morrinst/hgdp/faq.html (accessed 8 June 2002).

———. 1995. Proposed Model Ethical Protocol for Collecting DNA Samples—Draft (October).

———. 1997. "Proposed Model Ethical Protocol for Collecting DNA Samples." *Houston Law Review* 33: 1431–73.

Om, Michael and Howard Winant. 1994. *Racial Formation in the United States: from the 1960s to the 1990s* (New York and London: Routledge).

O'Toole, Kathleen. 1998. "Anthropology Department Likely to Split." Stanford Report (online) (20 May) www.stanford.edu/dept/new/report/news/may20/anthro520.html (accessed 5 October 2001).

Pascoe, Peggy. 1996. "Miscegenation Law, Court Cases, and Ideologies of 'Race' in Twentieth-Century America." *The Journal of American History* June: 44–69.

Paul, Diane. 1984. "Eugenics and the Left." *Journal of the History of Ideas* 45: 567–90.

———. 1995. *Controlling Human Heredity: 1865 to the Present* (New Jersey: Humanities Press).

Pennisi, Elizabeth. 1997. "NRC OKs Long-Delayed Survey of Human Genome Diversity." *Science* 278 (24 October).

Petit, Charles. 1998. "Trying to Study Tribes While Respecting Their Cultures." *San Francisco Chronicle* (19 February): A1.

Pinch, Trevor. 1986. *Confronting Nature: The Sociology of Solar-Neutrino Detection* (Dordrecht: Reidel).

Pollack, Andrew. 2003. "Large DNA File to Help Track Illness in Blacks." *New York Times* (27 May): A1.

Pomfret, John, and Deborah Nelson. 2000. "In Rural China, a Genetic Mother Lode." *Washington Post* (20 December): A1.

Potter, Elizabeth. 1993. "Gender and Epistemic Negotiation." In *Feminist Epistemologies*. Ed. by Elizabeth Potter and Linda Alcoff (New York: Routledge): 161–86.

Proctor, Robert. 1988. *Racial Hygiene: Medicine under the Nazis* (Cambridge, MA: Harvard University Press).

Provine, William B. 1973. "Geneticists and the Biology of Race Crossing." *Science* 182 (November): 790–96.

———. 1986. "Genetics and Race." *American Zoologist* 26: 857–87.

Provine, William, R. C. Lewontin, John A. Moore, and Bruce Wallace. 1981. *Dobzhansky's Genetics of Natural Populations* (New York: Columbia University Press).

Provine, William, and Ernst Mayr, eds. 1982. *The Evolutionary Sythesis: Perspectives on the Unification of Biology* (Cambridge: Harvard University Press).

Rabeharisoa, Vololona, and Michel Callon. 2004. "Patients and Scientists in French Muscular Dystrophy Research." In Sheila Jasanoff, ed., *States of Knowledge: The Co-Production of Science and Social Order* (London: Routledge).

Rabinow, Paul. 1996. "Fragmentation and Redemption in Late Modernity." In *Essays on Anthropology of Reason* (Princeton: Princeton University Press).

———. 1999. *French DNA: Trouble in Purgatory* (Chicago: University of Chicago Press).

———. 2002. "Midst Anthropology's Problems." *Cultural Anthropology* 17(2): 135–49.

Rabinow, Paul, and Gisli Palsson. 2001. "Inter-Mediate Reflections on the de-Code Controversy." Unpublished manuscript.

Rawls, John. 1971. *A Theory of Justice* (Cambridge, MA: MIT Press).

Reardon, Jenny. 1997. "Proteins That Matter." Paper presented to Science Studies Reading Group, Department of Science and Technology Studies, Cornell University (Fall).

Rensberger, Boyce. 1993. "Tracking the Parade of Mankind Via Clues in the Genetic Code." *Washington Post* (22 February): A3.

Risch, Neil, et al. 2002. "Categorization of Humans in Biomedical Research: Genes, Race and Disease." *Genome Biology* 3:2007.1–2007.12.

Robbins, Rebecca. 1992. "Self-Determination and Subordination: The Past, Present, and Future of American Indian Governance." In *The State of Native America*. Ed. by M. Annette Jaimes (Boston: South End Press).

Roberts, Leslie. 1991a. "Scientific Split over Sampling Strategy." *Science* 252 (21 June): 1615.

———. 1991b. "A Genetic Survey of Vanishing Peoples." *Science* 252 (21 June): 1614–17.

———. 1992a. "How to Sample the World's Genetic Diversity." *Science* 257 (28 August): 1204–1205.

———. 1992b. "Anthropologists Climb (Gingerly) on Board." *Science* 258 (20 November): 1300–1301.

Rothman, David. 1987. "Ethics and Human Experimentation: Henry Beecher Revisited." *New England Journal of Medicine* 317: 1195–99.

Rural Advancement Foundation International. 1993. "Patents, Indigenous Peoples, and Human Genetic Diversity." *RAFI Communique* (May).

———. 1993b. "The HGDP 'HitList' Circa 1993." http://www.rafi.org/misc.hgdplist.html (accessed 18 October 1999).

———. 1993c. Press Release (26 October 26).

Said, Edward. 1979. *Orientalism* (New York: Random House).

Sandel, Michael. 1996. *Democracy's Discontents: America in Search of a Public Philosophy* (Cambridge, MA: Harvard University Press).

Sandel, Michael, ed. 1984. *Liberalism and Its Critics* (New York: New York University Press).

Sanders, Douglas. 1980. *Background Information on the World Council of Indigenous Peoples* (Olympia, Washington: Center for World Indigenous Studies).

Sankaran, Neeraja. 1995. "African American Genome Mappers Pledge to Carry on Despite Grant Rejection," *The Scientist* 9(5): 1.

Schmidt, Charles W. 2001. "Spheres of Influence: Indi-gene-ous Conflicts." *Environmental Health Perspectives* 109: A216–A219.

Sclove, Richard E. 1995. *Democracy and Technology* (New York: Guilford Press).

Scott, Joan. 1992. "Evidence." In *Feminists Theorize the Political*. Ed. by Judith Butler and Joan Scott (New York and London: Routledge): 22–40.

Shapin, Steven. 1994. *A Social History of Truth* (Chicago: The University of Chicago Press).

Shapin, Steven, and Simon Schaffer. 1985. *Leviathan and the Air-Pump: Hobbes, Boyle, and the Experimental Life* (Princeton: Princeton University Press).

Sharp, Richard. 1998. "The Environmental Genome Project." A paper presented to the The First North American Conference on Genetics and Native Peoples, Polson, MT (11–12 October).

———. 2002. "How Informed Can Informed Consent Be?" Talk presented at the Human Genetics, Environment, and Communities of Color: Ethical and Social Implications Conference, West Harlem, New York (4 February).

Sharp, Richard, and Morris Foster. 2000. "Involving Study Populations in the Review of Genetic Research." *Journal of Law, Medicine and Ethics* 28(1): 41–51.

———. 2002. "Race, Ethnicity, and Genomics: Social Classifications as Proxies of Biological Heterogeneity." *Genome Research* 12: 844–50.

Shelton, Brett. 1998. "Genetic Research and Native Peoples." Indian Health Board Issue Briefing (18 February).

Shipman, Pat, and Alan Walker. 1996. *The Wisdom of the Bones: In Search of Human Origins* (New York: Alfred A. Knopf).

Shiva, Vandana. 1997. *Biopiracy: The Plunder of Nature and Knowledge* (Toronto: Between the Lines).

Shostak, Sara. 2003. *Locating Gene-Environment Interaction: Disciplinary Emergence in the Environmental Health Sciences, 1950–2000*. Ph.D. diss., University of San Francisco, California.

Silverman, Rachel. 2000. "The Blood Group 'Fad' in Post-War Racial Anthropology." In *Kroeber Anthropological Society Papers*. Ed. by Jonathan Marks (Berkeley: University of California Press): 11–27.

Simpson, George Gaylord. 1950. *The Meaning of Evolution: A Study of the History of Life and of Its Significance for Man* (New Haven: Yale University Press).

Skrentny, John. 2002. *The Minority Rights Revolution* (Cambridge, MA and London: Harvard University Press).

Smedley, Audrey. 1993. *Race in North America* (Boulder: Westview Press).

Smith, Anna Marie. 1994. *New Rights Discourse on Race and Sexuality: Britain, 1968–1990* (Cambridge: Cambridge University Press).

———. 1997. "The Displacement of Bio-Power with Pharmacological Technologies of Social Control." Talk presented to the Department of Science and Technology Studies, Cornell University.

———. 1998. *Laclau and Mouffe: The Radical Democratic Imaginary* (London and New York: Routledge).

Sober, Elliott. 1980. "Evolution, Population Thinking, and Essentialism." *Philosophy of Science* 37: 350–83.

Southern, David. 1971. *An American Dilemma Revisited: Myrdal's Study through a Quarter Century* (Ann Arbor, Michigan: University Microfilms).

Stanley, Barbara, and Michael Stanley. 1982. "Testing Competency in Psychiatric Patients." *IRB: A Review of Human Subjects Research* 4(8).

Stepan, Nancy Leys. 1982. *The Idea of Race in Science* (Hamden, CT: Archon Books).

———. 1986. "Race and Gender: The Role of Analogy in Science." *Isis* (77): 261–77.

Stocking, George W. Jr. 1982 (1968). *Race, Culture, and Evolution* (Chicago: The University of Chicago Press).

Stolberg, Sheryl Gay. 2001. "Ought We Do What We Can Do?" *New York Times* (12 August): D1.

Stoler, Ann. 1995. *Race and the Education of Desire: Foucault's History of Sexuality and the Colonial Order of Things* (Durham, NC: Duke University Press).

Stoneking, Mark. 1996. *Am. J. Hum. Genet.* 58: 595–608.

Sturr, Christopher J. 1998. *Ideology, Discursive Norms, and Rationality.* Ph.D. diss., Cornell University, Ithaca, New York.

Swedlund, Alan. 1993. "Is There an Echo In Here?: Historical Reflections on the Human Genome Diversity Project." Paper Presented to the Wenner-Gren conference, Anthropological Perspectives on the Human Genome Diversity Project (Sever Springs Center, Mt. Kisko, New York, 3–7 November).

Tallbear, Kimberly. 2003. "DNA, Blood, and Racializing the Tribe." *Wicazo Sa Review* 18(1): 81–107.

Tierney, Patrick. 2000. "A Fierce Anthropologist." *The New Yorker* (9 October): 50–61.

Turnbull, Colin M. 1962. *The Forest People* (New York: Simon and Schuster).

———. 1986. "Survival Factors among Mbuti and Other Hunters of the Equatorial African Rain Forest." In *African Pygmies*. Ed. by Luca Cavalli-Sforza (London: Academic Press).

UNESCO. 1952a. *The Race Concept: Results of an Inquiry* (Paris: UNESCO).

———. 1952b. *What is Race?* (Paris: UNESCO).

United Nations. 1949 (1948). *Universal Declaration of Human Rights* (Washington DC: U.S. Department of State).

Van Horne, Winston A. 1997. "Introduction." In *Global Convulsions: Race, Ethnicity, and Nationalism at the End of the Twentieth Century.* Ed. by Winston A. Van Horne (Albany: State University of New York Press): 1–45.

Venter, Craig. 2000. "Statement on Decoding of Genome." *New York Times* (27 June): D8.

Vizenor, Gerald. 1978. *Wordarrows: Indians and Whites in the New Fur Trade* (Minneapolis: University of Minnesota Press).

Wade, Nicholas. 1999. "DNA Backs a Tribe's Tradition of Early Descent from the Jews" *New York Times* (9 May): 1A.

Warren, Kay. 1998. *Indigenous Movements and Their Critics* (Princeton: Princeton University Press).

Weaver, Jace. 1997. *That the People Might Live: Natve American Literatures and Native American Community* (Oxford: Oxford University Press).

Weindling, Paul. 1989. *Health, Race and German Politics between National Unification and Nazism, 1870–1944* (Cambridge: Cambridge University Press).

Weiss, Kenneth M. 1993. "Letter Dated April 30, 1993, to Senator Akaka." In Committee on Governmental Affairs, *Human Genome Diversity Project: Hearing before the Committee on Governmental Affairs*, United States Senate, 103rd Congress 1st Session, 26 April: 43–44.

Weiss, Kenneth M., and Rick A. Kittles. 2003. "Race, Ancestry, and Genes: Implications for Defining Disease Risk." *Annu. Rev. Genomics Hum. Genet.* 4:33–67.

Weiss, Mark. 1991. "Saving Aboriginal DNA." *Science* 253 (27 September): 1467.

Weiss, Rick. 1995. "Academics Warn That Misunderstanding of Genetics Could Fuel Racism in U.S." *Washington Post* (20 February): A7.

Wexler, Alice. 1995. *Mapping Fate: A Memoir of Family, Risk, and Genetic Research* (Berkeley: University of California Press).

Whitt, Laurie Anne. 1998. "Biocolonialism and the Commodification of Knowledge." Paper presented to The First North American Conference on Genetics and Native Peoples, Polson, MT (October 11–12).

Wiercinski, Andrzej. 1964. "Comments." *Current Anthropology* 5(4): 314.

Wilkins, David. 1997. *American Indian Sovereignty and the U.S. Supreme Court* (Austin: University of Texas Press).

Williams, Raymond. 1983. *Keywords: A Vocabulary of Culture and Society* (New York: Oxford University Press).

Wilson, Allan. 1990. "Will Sequencing the Human Genome Revolutionize Biology?" *The New Biologist* (July): 585–86.

Wilson, James, et al. 2001. "Population Genetic Structure of Variable Drug Response." *Nature Genetics* 29(3): 265–69.

Wolpoff, Milford, and Rachel Caspar. 1997. *Race and Human Evolution: A Fatal Attraction* (Boulder, CO: Westview Press).

Wright, Lawrence. 1994. "The Illusion of Race in the United States." *Sacramento Bee* (28 August): F1.

———. 1994. "One Drop of Blood," *The New Yorker* (25 July): 46–55.

Wright, Robert. 1990. "Achilles' Helix." *New Republic* (19 July): 21–31.

Wynne, Brian. 1996. "Misunderstood Misunderstandings: Social Identities and Public Uptake of Science." In *Misunderstanding Science? The Public Reconstruction of Science and Technology*. Ed. by A. Irwin (Cambridge: Cambridge University Press).

Yona, Nokwisa. 2000. "DNA Testing, Vermont H. 809, and the First Nations" *Native American Village*. http://www.google.com/search?q=cache:205.25375.10/native/special/ht809200.html+Nokwisa+Yona%2BDNA&hl=en. (accessed 21 February 2001).

Zalewski, Daniel. 2000. "Anthropology Enters the Age of Cannibalism." *New York Times* (8 October). http://www.nytimes.com/2000/10/08/weekinreview/08ZALE.html (accessed 5 August 2002).

Zizek, Slavoj, ed. 1994. *Mapping Ideology* (London: Verso).

Index

Abuleas de Plaza de Mayo. *See* Grandmothers of the Plaza de Mayo
Acosta, Isidro, 111, 112
affirmative action policies, 13–14, 127, 132
Africa Genome Initiative, 207–8n. 250
African American, category of, 126, 128, 137–38, 139, 155, 163–64, 203n. 207, 204n. 219; and genomic research, 130–33, 151–52, 202n. 197. *See also* African Burial Ground Project; representation
African Burial Ground Project (ABGP), 130, 133–35, 136, 152, 164–65, 204n. 213
American Anthropological Association, 194–95n. 149
American Association for the Advancement of Science (AAAS), 50
American Association for Physical Anthropologists (AAPA), 83
American Indian Movement (AIM), 104
American Journal of Public Health, 153
Ames, Bruce, 50
Amnesty International, 51
"Answers to Frequently Asked Questions about the Human Genome Diversity Project" (NAmC), 153
anthropology, 43, 62, 66, 71, 86, 87, 93, 185n. 60, 192n. 129, 192–93n. 133; and colonialism, 89–90; expertise of, 80–82, 96; and historical relationship to biology (population genetics), 62, 85–87; and links with government agencies, 90, 194–95n. 149; struggles over the definition of, 83–92. *See also* biological anthropologists; cultural anthropologists; Diversity Project Anthropology Workshop; expertise, struggles over; physical anthropologists; race, and science, and anthropology; Wenner-Gren International Symposium
Anthropology Workshop. *See* Diversity Project Anthropology Workshop
Appiah, Kwame Anthony, 23
Argentina's Dirty War, 2, 49

Baldwin, Kate, 25
Barkan, Elazar, 22, 23, 24
Basic Call to Consciousness, A (Akwesasne Notes), 198n. 171
Bass, Thomas, 186n. 76
Bateson, Gregory, 58
Beecher, Henry, 115
Bell Curve, The (Hernstein and Murray), 53
Belmont Report, the, 116, 201n. 189
Bethesda ethics workshop (Ethical, Legal and Social Implications Program [ELSI]), 106, 111, 129, 197n. 167
biocolonialism, 15, 106, 138, 206n. 236
Biodiversity Convention, 111
bioethics, 197n. 163

biological anthropologists, 191n. 119
biology, 62, 71, 185n. 59, 186n. 71, 189n. 99, 205n. 229; and the category of population, 45; descriptive, 64; explanatory, 64; and physics, 188n. 95; and race, 12, 18, 31, 36, 38, 43–44, 89, 153–54. *See also* molecular biology
"Biology of Variation, The" (Johnson), 33
biomedical ethics, 122–24
biopower, 5, 75
bioprospecting, 15
"Bioscience under Totalitarianism Regimes" (Muller-Hill), 187n. 81
Birdsell, Joseph, 65–66
Blakey, Michael, 92, 93, 133–34, 135, 136, 138, 152, 204n. 219
blood groups, 42, 184n. 53, 185n. 57
Boas, Franz, 55–56, 90
Bodmer, Walter, 46, 69–71, 200n. 182
Bowcock, Anne, 71
Boyd, William, 41, 185n. 57
Brace, Loring C., 35–36, 37, 40
Branch, Taylor, 182n. 29
Brown v. Board of Education, 55
Bush, George H. W., 45
Buzzati-Traverso, Adriano, 63–64, 65, 66, 78

Canadian First Nations, 143, 144, 151, 160
Cantor, Charles, 50, 179n. 1
Castle, William E., 59
Caucasoid, category of, 72
Cavalli-Sforza, Luca, 4–5, 11, 46, 48, 50, 51, 53–54, 56, 64, 69–71, 72–73, 76–78, 81, 83, 92, 93–95, 107, 152–53, 179n. 1, 186–87n. 79, 190n. 114, 194n. 146; on behavioral traits, 196–97n. 161; debates with William Shockley, 2, 93, 188n. 84; expedition to Africa, 85, 193nn. 137, 139; and the "man on the street" concept, 53, 55
Cayuqueo, Nilo, 106
Centers for Disease Control, 111
Centre d'Etude du Polymorphisme Humaine (CEPH), 130, 157, 207–8n. 250
Chagnon, Napoleon, 194–95n. 149
Churchill, Ward, 198–99n. 175
Civil Rights Act (1866), 24
Civil Rights Movement, 182n. 29
Clark, W. E. Legros, 70
classification, 38–40, 43
Clay, Jason, 106, 107
clines, 34
Clinton, Bill, 153
Cold Spring Harbor Symposium (CSH / 1950), 46, 61, 62–69, 72–73, 81, 95
Cold War, the, 19, 45; views of race during, 25, 182nn. 28, 29
colonialism, 76, 89–90, 93, 145–46, 161, 194n. 147; and DNA patenting issues, 108–13; and North/South relations, 200n. 181. *See also* biocolonialism
"color-blind society," 55
community consultation, 166–67
community review, 201n. 194
Cook-Deegan, Robert, 2, 47–49, 53, 179n. 1, 186nn. 75, 78, 186–87n. 79
Coon, Carleton, 31, 35, 42, 86, 184n. 47, 188n. 88
co-production, 6–10, 101, 152, 161
covertness criterion, 180–81n. 8
cultural anthropologists, 55–56, 85–86, 89; and points of difference between cultural and physical anthropology, 90–92
Cultural Survival Enterprises, Inc., 106, 194n. 141
culture, concept of, 20–22, 32, 88–89
Current Anthropology, 34
Cuvier, Georges, 20

Darkness in El Dorado (Tierney), 194–95n. 149
Darwin, Charles, 33, 57–58, 67
Darwinian variation, 33
Davenport, Charles, 24
Decade of Indigenous Peoples, 198n. 174
Demerec, Milosav, 62, 189n. 101
Democracy's Discontents (Sandel), 123
deoxyribose nucleic acid (DNA), 54, 129, 141, 187n. 80; mitochondrial, 191–92n. 121; patenting of, 110; sampling, 142–44, 158, 164; testing, 140, 205n. 229
Department of Energy (DOE), 1–2
Diamond, Jared, 50
DiRienzo, Anna, 186n. 72
"disappearing." *See* "vanishing," rhetoric of
discourse, 181n. 9; political character of, 180–81n. 8
Diversity Project. *See* Human Genome Diversity Project
Diversity Project Anthropology Workshop (Penn State University / 1992), 81–82, 84, 88, 100, 101, 194n. 144, 197n. 168

DNA Polymorphism Discovery Resource (PDR), 157
Dobzhansky, Theodosius, 11, 12, 21, 24, 34–36, 37, 53, 56, 64, 67–68, 185n. 58; views on classification, 38–40; views on pure race, 183n. 36, 184nn. 48, 49, 184–85n. 54
Du Bois, W.E.B., 25
Dukepoo, Frank, 137–38, 144–45, 149, 205n. 221
Dulbecco, Renato, 186–87n. 79
Dunn, Leslie Clarence, 24, 36–37, 39, 41, 183n. 35, 184nn. 52, 53
Dunston, Georgia, 130–31, 132, 138, 151, 152, 163–64, 203nn. 207, 208

Emancipation Proclamation (1863), 24
emergence, 8. *See also* co-production
"empty matrix problem," 80–81. *See also* sampling strategies
Engels, Frederick, 181n. 10
Enlightenment, the, 7–8, 180–81n. 8
Environmental Genome Project, 146, 208nn. 251, 252
ethical, legal and social implications (ELSI), 106, 111, 129, 187n. 83
ethics: entanglement with science, 100–101, 115, 118, 161–62; as expertise, 100; limits of, 122–25; and research protocols, 114–15. *See also* Model Ethical Protocol; subjects, formation of
ethnic group, concept of, 36, 77, 126. *See also* population approach to race
eugenics, 12, 24, 32, 42, 53; "mainline eugenics," 60; new eugenics, 15, 187n. 81
evolution, 33, 41–42, 43, 56–57, 59, 64; "blending" theory of, 188n. 94; and genetics, 62–63, 188n. 96; and genome diversity, 131; and molecular traits of human populations, 70; and physical traits in the study of, 55, 188n. 88; and race formation, 60–61. *See also* fossils
Evolutionary Synthesis, The (Mayr and Provine), 57
expertise: critical studies of, 74; integration of biological and cultural, barriers to, 87–88; material constituents of, 87; and power and authority, 4–5, 75–76, 96, 125; struggles over, 10, 78–81, 83–87. *See also* United Nations Educational, Scientific and Cultural Organization, and "experts on race problems"

false consciousness, 181n. 10
Feldman, Marcus, 81, 114, 194n. 146
Fields, Barbara, 23
First North American Conference on Genetic Research and Native Peoples, 145–46, 161–62
Fisher, Ronald, 31, 59, 60
Fossil Evidence for Human Evolution, The (Clark), 70
fossils, 41, 53, 64–65, 190n. 113
Foucault, Michel, 5, 181n. 15; and the power-knowledge relationship, 6, 75, 191n. 120
Freeman, Harold, 153
Fried, Morton, 94

Gannett, Lisa, 37, 183n. 45, 184n. 48
Gates, Henry Louis, 23
gel electrophoresis, 54
General Agreement on Tariffs and Trade (GATT), 111
General Services Administration (GSA), 133
Genes, Peoples and Languages (Cavalli-Sforza), 190n. 114
geneticists, 41, 43, 44, 45, 47, 54, 59, 64–65, 79, 83, 84, 190n. 116, 193n. 136; criticisms of by anthropologists, 189n. 102; and physical traits in the study of evolution, 55, 188n. 88; sampling procedures of, 51–52
genetics, 42, 59, 86; mathematical, 189n. 98; and the study of evolution, 62–63, 188n. 96. *See also* population genetics
Genetics of Human Population, The (Cavalli-Sforza and Bodmer), 46, 64, 69, 71, 72, 189–90n. 108
Genetics and the Origin of Species (Dobzhansky), 12, 64
"Genetics and Race" (Provine), 185n. 62
Genetics and the Races of Man (Boyd), 185n. 57
Genomic Research in African-American Pedigrees (G-RAP) project, 130–31, 132, 152, 203n. 208
Genomics, 45, 51, 52, 74, 79–80, 152, 153, 179, 186–87n. 79
Goldschmidt, Richard, 59
Goodman, Allan, 87, 194n. 143
governance, 15, 75, 158
Grandmothers of the Plaza de Mayo, 2, 49–50, 51

Greely, Henry T., 106–7, 108, 109, 112, 114, 199n. 176, 206n. 239; and the Model Ethical Protocol, 200–201n. 186
group consent, 99–100, 116–18, 127, 152, 163, 201n. 190
groups, 122–25, 189n. 107, 198–99n. 175; definitions of, 119–22, 163, 201n. 193; existence of, 118–19; isolated, 67–69; in nature, 152–54, 161. *See also* group consent

Haldane, J.B.S., 21, 59, 60
Handlin, Oscar, 19
Hansen, Roger, 200n. 181
Hernstein, Richard, 53
Higginbotham, Evelyn B., 23
History and Geography of the Human Gene (Cavalli-Sforza), 50, 153–54, 196–97n. 161
human beings, defining, 16, 76–78, 96; debates concerning, 78–82
human evolution. *See* evolution
human genetic diversity, 11–12, 35, 41–42, 49, 185n. 57; and Eurocentrism, 11, 72; as an object, 10–12, 162. *See also* human genetic diversity, major conferences
human genetic diversity, major conferences, 142; First International Conference on DNA Sampling (Montreal), 142–44, 206n. 236; Genetic Research and Native Peoples: Colonialism through Biopiracy (Flathead Reservation, Montana), 145–46; Stanford conference (Palo Alto), 144–45
Human Genome Diversity: A Proposal for Two Planning Workshops and a Conference (Cavalli-Sforza, Feldman, King, and K. Weiss), 81
Human Genome Diversity Project (HGDP), 2, 3–4, 11, 15–16, 19, 44, 60, 73, 96–97, 157–58, 166–67, 179n. 2, 204n. 216, 206n. 232, 208n. 262; and accusations of racism, 2, 3, 4–6, 24, 92–96, 158, 195n. 153; and accusations of threatening indigenous groups by its actions, 98–99, 122–25, 205n. 228, 206n. 238; and classification, 39–40; debates concerning, 13, 32, 180n. 5, 186n. 66, 198n. 171; different understandings of, 158–60; and disease research, 141–42, 204n. 211, 206n. 233; funding of, 186n. 71, 186–87n. 79; inclusion problems, 160–62 (*see also* Human Genome Diversity Project, and participation); and the indigenous rights movement, 103–8; 199nn. 178, 179; initial proposal for, 47–49, 68; involvement of anthropologists in, 78–87, 192n. 129; and "major ethnic groups," 5, 92, 132; origins, 45–46; and patenting, 108–12; and race formation, 71–72; related projects, 208nn. 251, 262; and simultaneous emergence, 8–9; as the "Vampire Project," 2, 98, 158. *See also* African American, category of; Diversity Project Anthropology Workshop; human genetic diversity; Human Genome Diversity Project, ethical concerns about; Human Genome Diversity Project, and participation; Model Ethical Protocol; Native American, category of; North American Regional Committee; Wenner-Gren International Symposium
Human Genome Diversity Project, ethical concerns about, 100–103, 187n. 81, 197nn. 164, 165, 167; and the indigenous rights movement, 103–6; and the problem of information, 106–8. *See also* ethics
Human Genome Diversity Project, and participation, 149–50, 153–56; beneficiaries of participation, 139–42; discourses of participation, 128–29; and the importance of history, 136–39; and the inadvertent construction of groups in nature, 152–54; and instrumentalism, 134–36; and local participation, 202–3n. 202; models of, 130; and objectification, 146–49; and participatory democracy, 126–28. *See also* African Burial Ground Project; Genomic Research in African-American Pedigrees (G-RAP) project; participation
Human Genome Organization (HUGO), 1, 105, 200n. 182
Human Genome Project (HGP), 1, 11, 15, 46, 47–49, 63, 107, 110, 130, 138, 186–87n. 79; and African Americans, 151; cost of, 110; ethical issues, 187n. 83; and fears of race hygiene and eugenics, 52–53
human groups. *See* groups
humanitarianism, market for, 45
human population genetics. *See* population genetics

human sciences: and power, 5, 75; definition of, 5, 75–76; fractures within, 83
Hunt, Edward, 67, 69

Iceland, genetic database controversy in, 181n. 15
Idea of Race in Science, The (Stepan), 21
identity. *See* subjects, formation of
identity politics, 14
ideology: concept of, 7–8, 165, 180–81n. 8, 190–91n. 117; and "functional false consciousness," 181n. 10. *See also* classification; knowledge, theories that oppose ideology
inclusion. *See* participation; voice, construction of
Indigenous Peoples Council on Biocolonialism, 205n. 228
Indigenous Rights Movement, 103–6, 198n. 71, 199nn. 178, 179, 200n. 184. *See also* World Council of Indigenous Peoples
informed consent, 161, 201nn. 187, 189, 190
Institute for the Study of Human Variation (Columbia University), 36–37, 184n. 53
International Haplotype Map (HapMap) Project, 4, 208n. 251; related projects, 208n. 251
International Indian Treaty Council (IITC), 104
"Is the Human Genome Diversity Project a Racist Enterprise?" (Moore), 94
isolated human populations, 67–69, 88, 128, 152–53, 198n. 173
isolates. *See* isolated human populations
"Isolates of Historical Interest," 68–69, 105, 198n. 173

Jackson, Fatimah Linda Collier, 131, 194n. 143
Jefferson, Thomas, 7
Jennings, Herbert Spencer, 24
Jensen, Arthur, 187n. 82
Jim Crow laws, 25
Johnson, Francis E., 33
Journal of the American Medical Association (JAMA), 47

Kant, Immanuel, 123
Kevles, Daniel, 60, 187n. 81
King, Martin Luther, 182n. 29
King, Mary-Claire, 2, 49–51, 81, 100, 129, 179n. 1, 186nn. 72, 75, 76
Kitcher, Phillip, 202n. 201
Kluger, Richard, 25
knowledge: relationship to institutions, 6, 164–66; relationship to power and social order, 75, 127, 164–66; theories that oppose ideology, 7–8, 12, 25–29, 32–33, 194n. 148, 161. *See also* Foucault, Michel, and the power-knowledge relationship
Kuklick, Henrika, 90
Kurds, the, 48

Laclau, Ernsto, 123
Laughlin, Harry H., 24
Laughlin, W. S., 67
Lewontin, Richard, 35, 181n. 15, 184n. 50
Lincoln, Abraham, 24
Livingstone, Frank, 23, 33, 34–36, 37, 38, 53
Locke, John, 123
lynchings, 25, 42

MacArthur Foundation (The John D. and Catherine T. MacArthur Foundation), 114–15, 139, 140–41, 144, 149, 200n. 184, 202n. 200
Manifesto on Genomic Studies among African Americans, 151–52
Mannheim, Karl, 180–81n. 8
Man's Most Dangerous Myth (Montagu), 36
Maori, representation of, 120
Marks, Jonathan, 192n. 129
Marxism, 7
Mather, Kenneth, 30–31
Maybury-Lewis, David, 86, 194n. 141
Mayr, Ernst, 21, 57, 58–60, 188n. 96
Mendel, Gregor Johann, 54, 57
Mendelian genetics, 57–60. *See also* population genetics, emergence of
Mendel's laws, 63, 64, 70
Mexican American, as biomedical category, 132
"micro-evolution," 37
Micronesia, 67
migration theory, 95, 196n. 160
miscegenation, 71
mitochondrial Eve study, 191n. 121
Mittman, Ilana Suez, 131

Model Ethical Protocol (MEP), 99, 114–16, 145, 146–47, 161, 163, 200–201n. 185, 201n. 192; and the "Partnership with Participating Populations" program, 126–27, 134, 135–36, 152
molecular biology, 55, 63
Momaday, Scott, 140
Montagu, Ashley, 26, 36, 42, 56, 66, 94–95, 182n. 31; use of the term "ethnic group," 36, 188–89n. 97; views on race, 184n. 51
Mooney, Pat, 106–7, 109, 11, 112
Moore, John, 93, 94–95, 116, 195n. 155, 196n. 158, 196–97n. 161; on migration theory, 196n. 160
Morgan, Thomas Hunt, 58, 183n. 41
Mouffe, Chantal, 123
Muller, Herman J., 30, 183n. 41, 187n. 80
Muller-Hill, Benno, 187n. 81
Murray, Charles, 53
Myrdal, Gunnar, 25

National Center for Human Genome Research (NCHGR), 53
National Congress of American Indians, 157
National Genetic Data Bank (Argentina), 186n. 73
National Human Genome Research Center (NHGRC), 1
National Human Genome Research Institute (NHGRI), 4, 157, 158–59, 166, 208n. 254
National Institute of General Medical Sciences (NIGMS), 1; and the Human Genetic Cell Repository, 166
National Institutes of Health (NIH), 11, 47, 71, 81, 101, 111, 146, 166, 180n. 4, 197n. 164; and the NIH Revitalization Act (1993), 153
National Research Council, 145
National Science Foundation (NSF), 1; Physical Anthropology Program, 79–80
Native American, category of, 14–15, 68, 139–40, 155, 163, 205n. 222. *See also* Vanishing Indian myth
Native-L, 105
natural selection, 40, 58, 189n. 99
Nazism, and race, 17, 24, 27, 28, 29, 32, 38, 42; criticisms of Nazi racial practices, 30–31
Neel, James, 67, 85–86, 193–94n. 140, 195n. 149
Negrito, category of, 72
Negroid, use of the term, 190–91n. 117
"Negro problem," the, 24–25, 182n. 26; Carnegie Commission study of, 25
New Scientist, 92–93
"new world order," 45, 47
North American Regional Committee (NAmC), 99, 114–16, 124–25, 127, 132, 134, 145, 153, 165, 196n. 158; and group consent, 116; and groups as the unit of consent, 116–18. *See also* groups
Notion of the Tribe, The (Fried), 94

Omi, Michael, 188n. 90
"On the Non-Existence of Human Races" (Livingstone), 53
On the Origin of Species (Darwin), 57
Origin and the Evolution of Man (Cold Spring Harbor Symposium), 46
Origin of Races, The (Coon), 35
Out-of-Africa theory, 139, 205n. 227
"Out of Eugenics: The Historical Politics of the Human Gene" (Kelves), 187n. 81

Paabo, Svante, 186n. 72
participation: and epistemology, 152–55; as ethical practice, 126–27, 129, 134; history of in Western liberal democracies, 127, 130, 136–38; and inclusion/exclusion, 142–46, 160–62; as material practice, 142–46; as political strategy, 134–36; problems with, 139–42, 146–48, 154–55, 161; of racial and ethnic groups in genetic research, 4, 130–33, 154. *See also* African Burial Ground Project; Genomic Research in African-American Pedigrees (G-RAP) project; Human Genome Diversity Project, and participation
Pascoe, Peggy, 55
patenting: accusations of, 2, 105–6; activist opposition to, 102, 109; debate over relevance to Diversity Project, 108–12; and Guaymi case, 109–12
phenotype, 54; and phenotype-genotype distinction, 40–42; and race formation, 185n. 57
physical anthropologists, 29, 30, 32, 41–44, 54, 71, 163, 192nn. 125, 130, 193n. 136; and anthropometric studies of brain cases, 69–70; as defined by Dunn, 41; and the Human Genome Diversity Project, 74–76, 79–80; and points of dif-

ference between cultural and physical anthropology, 90–92. *See also* anthropology; race
Physicians for Human Rights, 51
physics, 188n. 95
Piazza, Alberto, 95
pleiotropy, 60
polygenism, 182n. 20
polygeny, 60
population: as conceptual novelty in genetics, 63–64, 192n. 124; demographer's concept of, 64; Mendelian concept of, 64; as replacement for race, 5, 8, 12, 21–22, 32, 35–36, 66, 153. *See also* population approach to race; population genetics; race, population concept of
population approach to race, 32–33, 37, 38, 40, 43, 65–66, 163, 190n. 109, 204n. 218
population genetics, 12, 41, 42, 45, 46, 56, 71, 185n. 58, 188n. 96; emergence of, 57–60; and natural population, 63; and race, 60–61, 69–71; as a science of liberation, 50, 181n. 18, 186n. 74
power: modern formations of, 75; theories of, 3, 7–8. *See also* science, and relation to power; expertise; human sciences
power-knowledge. *See* Foucault, Michel
Primer and Resource Guide (Indigenous People's Council on Biocolonialism), 140
privatization, problem of, 166
Project Camelot, 90
Provine, William, 185n. 62
Pygmies, 70, 85, 193n. 139; and their relationships with farmers, 193n. 137

Rabinow, Paul, 181nn. 15, 17
race, 4–5, 36–37, 163, 188n. 90; bio-historical concept of, 40; categorization of, 38–40; definition by geneticists and physical anthropologists in Second UNESCO Statement, 30–32; definition by sociologists in First UNESCO Statement, 27–30; definition of pure race, 183n. 36; genetic definition of, 46, 69–70, 72; gradation, of, 21; and ideology, 18–19, 21; and IQ, 53, 187n. 82; "man on the street" construction of, 53–54; as molecular and "invisible," 54–55; narrative of the decline of, 17, 23–24; as obsolete concept of the life sciences, 8–10; physical concept of, 55–56, 73; population concept of, 40, 61, 183n. 38; as site of differences among physical anthropologists and geneticists, 36–42, 73, 93–96, 159–60; typologies of, 22–23. *See also* biology, and race; population genetics, and race; race, and population/typological distinction; race, and science; UNESCO Statements on Race
race, and population/typological distinction, 32–34, 183n. 45; and classification, 38–40; cultural approach, 36–38; dichotomies distinguishing a typological approach from a population approach, 33; and phenotype/genotype distinction, 40–42; population approaches, 34–36, 43
race, and science, 4, 9–16, 17–19, 32, 34–35, 43–45, 53–56, 153–54, 181n. 11, 183n. 43, 188n. 90; and anthropology, 20, 23, 24; histories of, 19–23. *See also* biology, and race
Race, Culture and Education (Stocking), 19–20, 88
Race and Nationality in American Life (Handlin), 19–20
Racial Formation in the United States (Omi and Winant), 188n. 90
racial qualifiers, 202nn. 197, 198, 203n. 206
racial variations, 62–63
racism, 10, 25–26, 76, 93, 94, 161. *See also* Human Genome Diversity Project, and accusations of racism; "Negro problem," the
Rawls, John, 123
Reid, Walt, 106, 109
representation: of "African Americans," 130–32, 151–52, 165; of "communities," 166–67, 201n. 194; of the Diversity Project, 107–8; of indigenous peoples, 106–7, 112–13; and objectivity, 107–13; political content of, 113; as political strategy, 149–50. *See also* African American, category of; Native American, category of; participation; voice
research: and informed consent, 201n. 187; regulation of, 115
Retreat of Scientific Racism, The (Barkan), 22
Rifkin, Jeremy, 197n. 164
Roberts, Leslie, 51, 192n. 126
Rousseau, Jean-Jacques, 7
Rural Advancement Foundation International (RAFI), 98, 99, 102–3, 108–12

sampling strategies: empty matrix problem, 80–81; population versus grid approach, 76–78; and representation, 82
Sandel, Michael, 123, 201n. 195
Santa Clara Pueblo v. Martinez, 121
Sawyer hearings, the, 207n. 248
science: and democracy, 202n. 201; and ideology, 18–19; and informed consent, 161, 201nn. 187, 189, 190; ; and politics, 24; and racial qualifiers, 202nn. 197, 198; and relation to power, 3; and the "science first" approach, 79, 192n. 128. *See also* race, and science
Science, 74, 79, 129
science and technology (S&T), 6–7, 180n. 7
segregation laws. *See* Jim Crow laws
Shockley, William, 2, 93, 188n. 84
Silverman, Rachel, 41
Silverman, Sydel, 195n. 154
Simpson, George Gaylord, 64–65
Skrentny, John, 25
Snyder, Laurence, 66–67
social constructivism, 9, 21, 165
social determinism, 7, 165
Southern, David, 25
Sponsel, Leslie, 194–95n. 149
Stanford University, 91
Statements on Race. *See* UNESCO Statements on Race
Stepan, Nancy, 21, 22, 23, 182n. 21; and the "new science of human diversity," 21
Stocking, George, 19, 21, 88, 182n. 20; definition of race, 20
Stoneking, Mark, 186n. 72
Stover, Eric, 49–50
Strandskov, Herluf, 62–63
"stupidity," 20
Sturr, Christopher, 180n. 8
Sturtevant, Alfred H., 59
subjects, formation of, 9, 99, 103, 117–20; and ethical protocols, 117–20, 161; and object formation, 152–54, 162–64; and social power, 120–24; theory of "looping," 9
Swedlund, Alan, 92–94, 194n. 143, 195–96n. 157
symmetry, 7

Tallbear, Kimberly, 205n. 229
Thieme, Frederick, 67
Third World Network, 2, 98, 105, 108, 113
Tierney, Patrick, 194–95n. 149
tribe, category of, 92, 94
truth: Enlightenment concept of, 7; relationship to ideology, 7–9. *See also* ideology
Turkey, 48
Turnbull, Colin, 85, 193n. 139
Turner, Terry, 194–95n. 149
Tuskegee Syphilis Experiment, 139
typology. *See* race, and population/typological distinction; race, typologies of

UNESCO Statements on Race, 12, 19, 23–26, 66, 181n. 13; First Statement, 26–29, 32, 53, 189n. 101; Second Statement, 29–31, 53–54, 183n. 36
United Nations, 104
United Nations Convention on Biological Diversity (CBD), 143
United Nations Economic and Social Council, 27
United Nations Educational, Scientific and Cultural Organization (UNESCO), 12, 26–27; and "experts on race problems," 26, 30. *See also* UNESCO Statements on Race
United Nations Food and Agriculture Organization (FAO), 102
Universal Declaration of Human Rights, 13, 104
universal humanism, doctrine of, 13, 104–5
University Federation for Democracy and Intellectual Freedom (UFDIF), 24
U.S. Department of Energy (DOE), 1–2, 11, 47, 81
U.S. immigration policy, 24

Valencia genome meeting, 49–50
"Vampire Project," 2, 98, 156
Van Horne, Winston A., 23–24
"vanishing," rhetoric of, 1, 82, 105, 148
Vanishing Indian myth, 198n. 172
voice, construction of, 149–52, 165. *See also* participation; representation
Voltaire, François-Marie, 7

Washburn, Sherwood, 192n. 30
Watson, James, 47, 48, 179n. 1, 186n. 72, 187nn. 81, 83
Weaver, Jace, 198n. 172

Weinberg, Wilhelm, 189n. 98
Weiss, Kenneth M., 81, 82, 86, 96, 131–32, 204nn. 211, 212
Weiss, Mark, 74, 79–81, 82, 95, 96
Wenner-Gren Foundation, 194n. 143, 195n. 144
Wenner-Gren International Symposium (Cabo San Lucas, Mexico), 87–89, 93, 133, 195n. 156
Wexler, Nancy, 51
What is Race? (UNESCO), 28
Wiercinski, Andrzej, 39, 41, 42
Wilson, Allan, 63, 76–78, 179n. 1
Winant, Howard, 188n. 90
"Winds of Death" (Cook-Deegan), 48
World Congress of Indigenous Peoples, 2
World Council of Indigenous Peoples (WCIP), 104, 106, 107, 113, 151, 161, 199n. 197
World Court, the, 104
World Resources Institute, 106
World War II, 12, 19, 42, 104
Wright, Sewall, 21

Yanagisako, Sylvia, 91
Yanomami people (Brazil), 67, 86, 194–95n. 149
Yarmolinsky, Michael, 63

Zizek, Slavoj, 190–91n. 117

The Shadows and Lights of Waco: Millennialism Today
by James D. Faubion

Life Exposed: Biological Citizens after Chernobyl
by Adriana Petryna

Anthropos Today: Reflections on Modern Equipment
by Paul Rabinow

When Nature Goes Public: The Making and Unmaking of Bioprospecting in Mexico
by Cori Hayden

Picturing Personhood: Brain Scans and Biomedical Identity
by Joseph Dumit

Fiscal Disobedience: An Anthropology of Economic Regulation in Central Africa
by Janet Roitman

Race to the Finish: Identity and Governance in an Age of Genomics
by Jenny Reardon